THE PATTERNS OF
NEW IDEAS

THE PATTERNS OF NEW IDEAS

✦

300 Ideas for Products, Inventions and Improvements

Mark Meek

iUniverse, Inc.
New York Lincoln Shanghai

THE PATTERNS OF NEW IDEAS
300 Ideas for Products, Inventions and Improvements

iUniverse, Inc.

For information address:
iUniverse, Inc.
2021 Pine Lake Road, Suite 100
Lincoln, NE 68512
www.iuniverse.com

ISBN: 0-595-31163-6

Printed in the United States of America

Contents

DEDICATION

I would like to dedicate this book to the country that brought me into the world.

At the time of this writing, English is the language of the world. A British pound is worth more than any other currency on earth. Soccer, a British development, is the world's most popular sport by far. A poll of Americans selected Tony Blair as the number one leader on earth, ahead of even George Bush. Almost certainly, more money flows through London's financial district than anywhere else in the world.

Harry Potter rules the literary world (whether or not one agrees with the spiritual implications of the stories). William Shakespeare is considered to have been the greatest writer in any language ever and his themes can be found all over modern television.

The BBC is easily the world's number one news service. The Titanic is the most famous ship in history. Even if you do not like rock music; The Beatles, The Rolling Stones, The Who and, Led Zeppelin really set the pace. Norman Baden Powell started the international Boy Scouts.

When it comes to scientific progress, Sir Edmund Halley discovered the most famous comet, Halley's comet. In 1808, John Dalton published "A New System of Chemical Philosophy" which marked the beginning of the modern science of chemistry, without which the modern world would be impossible. It is Oxford's Stephen Hawking that is today leading the search in Physics for the "Theory of Everything" and is the world's best-known physicist.

In 1825, Michael Faraday discovered benzene, which is the vital starting point of much of organic chemistry. Hydrogen is the most prominent element in the universe and was found by Henry Cavendish. Oxygen was found by Joseph Priestley and nitrogen was found by Daniel Rutherford. In 1939, Imperial Chemicals first produced the Polyethylene plastic that could be considered the beginning of the plastics revolution. Matches were invented by chemist John Walker in 1837.

When it comes to atoms; J.J. Thomson discovered the electron. James Chadwick discovered the neutron. The Russian, Mendeleev had come up with the idea of the periodic table of the elements but it was Henry Moseley that developed the idea of atomic numbers, thus straightening out the table, in 1914. The first nuclear accelerator was used by Cockcroft and Walton at Cambridge in 1932.

As for biology and optics, in 1267, Roger Bacon invented the magnifying glass. It was Robert Hooke who used the microscope for scientific purposes and who first described a biological cell. Insulin, a hormone, was the first protein to have it's structure worked out, it was accomplished by biochemist Frederick Sanger. It was Francis Crick, with American James Watson, that made the monumental discovery of DNA. At the time of this writing, Britain is far ahead of the rest of the world in DNA science.

It is the Italian Galileo that is given credit for inventing the refracting telescope. But it suffered from chromatic aberration until John Dolland invented the achromatic lens in 1758. Anyway, Isaac Newton's reflecting telescope can be built much larger than a refractor and is used in almost all major observatories today.

It was, of course, Isaac Newton who founded modern physics with his laws of motion and his development of calculus. It is Newton who first defined gravity, the force that rules the universe on a large scale. He also began our understanding of light by breaking it down into a spectrum with a prism.

Every time you see a photograph, remember William Henry Fox Talbot. The Frenchman Louis Daguerre came up with a method of taking photographs but it involved silver-plated copper. It was Talbot who made the practical paper prints and opened the first photography business in Reading as well as writing the world's first photographically illustrated book.

The names that most people associate with the development of movies are Thomas Edison and the French Lumiere brothers. However, the fact is that Eadweard Muybridge filmed a galloping horse by rapid photography to settle a bet between two men in England as to whether a horse's feet all leave the ground at once when galloping. This was the first film as it came before either Edison or the Lumieres made films.

In 1926, the first television was demonstrated in a London department store, even though it was a mechanical television and was not of the final design that came into use. In 1936, Isaac Schoenberg's team much improved television technology, using a 405-line image.

When it comes to electricity, William Sturgeon invented the electromagnet in 1823. Without this and it's application to motors, modern appliances would be impossible. Italy's Allesandro Volta first produced an electric current, from a chemical reaction, but this could not supply electricity on a large scale. The electromagnet is the basis for the generation of electricity.

Most of the world's electricity is generated using Charles Parson's steam turbine, regardless of how the steam is produced. The steam turbine also made possible much faster ships. Michael Faraday invented the capacitor and transformer, without which electrical power as we know it today would be unimaginable.

The first successful electric telegraph was patented in 1837 by William Cooke and Charles Wheatstone. This was a vital step in telecommunications.

In electronics, it was John Ambrose Fleming that invented the vacuum tube that made popular radio and television possible. The Italian, Marconi is generally given credit for development of the concept of radio. But it was done in England and subsidized by the British government when Marconi's own government showed no interest in his ideas.

America recognizes Thomas Edison as the inventor of the light bulb however, Britain recognizes it's own Joseph Swan. At any rate, it was Humphry Davy who started the race for a practical light bulb by making a strip of charcoal glow by passing an electric current through it in 1809.

Modern flying would not be possible without Radar, which was first used by Sir Edward Appleton in 1924. RAF officer Frank Whittle patented the jet engine in 1930. The only successful supersonic transport was the Concorde. The Harrier was the world's first practical vertical takeoff and landing plane. Christopher Cockerell invented the hovercraft. Rockets have been around for hundreds of years but the first step toward modern rockets since then was the addition of fins and a steering mechanism to an army missile by William Hale in the 1890s. Americans idolized Charles Lindbergh as the first to fly solo across the Atlantic but the fact is that two Englishmen preceded him, taking turns at the controls of the plane.

Early aircraft were notoriously unstable in flight. The Sopwith Camel was an early British fighter plane. It's designers found that if the wings were slanted slightly upward, when the plane went into a roll the wing on the way down would generate more lift than the wing on the way up. This counteracted the roll and gave the plane more stability. They also noticed that the rudder of the aircraft worked much better if it was not in the same horizontal plane as the wings. These innovations have been with airplanes ever since. The Sopwith Camel was such a stable and maneuverable aircraft that it's pilots could often make their German rivals in World War One aerial dogfights crash simply by drawing them into maneuvers that their planes could not handle.

The system of latitude and longitude that enables us to navigate the globe was developed in England and made practical by John Harrison's chronometer. The prime meridian, 0 degrees longitude, passes through Greenwich, a London suburb. Every time you look at the clock, remember that Greenwich Mean Time, GMT, was developed with latitude and longitude and is the standard of time in the world. In 1955, Louis Essen and Jack Parry developed the world's first atomic clock.

The slide rule, around which modern engineering was to revolve until electronic calculators came out, was invented by William Oughtred in 1622. Charles Babbage was considered to be the "father of the computer" with his mechanical computer. British Mathematician Alan Turing was the first to build a truly programmable calculator. It was Tim Berners Lee who invented the World Wide Web, without which widespread internet use would not be practical.

The Industrial Revolution, the beginning of the modern world, began in Manchester. Thomas Newcomen's steam engine was improved upon by James Watt and made it possible. In 1856, a way of mass-producing steel was developed by Henry Bessemer, without which the modern world would be unimaginable.

Josiah Wedgwood revolutionized the mass production of useful pottery items. Electroplating was patented by brothers George and Henry Elkington in 1840. Potassium was discovered by Humphry Davy, who isolated it from potash in 1807 in the first instance of metal isolated by electrolysis. In 1913, Harry Brearley developed stainless steel.

The railroad locomotive was invented by George and Robert Stephenson. In 1804, Richard Trevithick built the first successful steam locomotive. Can you imagine the settling of America without trains? In the 1950's Eric Laithwaite pioneered maglev trains. The first metal bridge was the "Iron Bridge" in England. London showed the world the first modern subway and arguably the world's greatest feat of engineering is the Chunnel.

England did not develop the automobile per se. But Fredrick Lanchester invented disc brakes in 1902. Oliver Joseph Lodge invented spark plugs. The pneumatic tire was first patented in Britain and Henry Harrison invented the automobile radiator. The world still considers the Rolls Royce and the Jaguar as the ultimate cars.

The concept of paving roads originated in England. The battle tank, around which modern warfare revolves, was introduced by Britain in World War One. Did you know that it was British Engineers that found the oil in the Middle East, which is so important to the world today? It was British Petroleum that found the large oilfield in Alaska.

The first commercially successful bicycle was built in 1885 by John Starley. Germany's Gottlieb Daimler invented the two-wheeled motorcycle but it was preceded by a year by Edward Butler's 1894 motorized tricycle.

The modern concept of insurance underwriting originated with Lloyd's of London. The first check was handled by the bankers Clayton and Morris in 1659. Postage stamps were introduced by Rowland Hill in 1840. The first ATM appeared in London in 1967.

When it comes to freedom, the modern idea of freedom definitely began with the Magna Charta. The U.S. constitution and the French 'The Rights of Man' are it's descendants. The modern concept of political right and left began when members of Britain's parliament would sit on either the right or left according to their political views.

The next time you put on your clothes remember James Hargreaves, who invented the spinning jenny in 1767 and John Kay, who invented the flying shuttle to make textiles rapidly. When you buy clothes in a color that you like, remember William Henry Perkin, who in 1856 created mauve, the first synthetic dye.

The next time your wife or parents make you cut the lawn instead of watching sports on the weekend, you can blame Edwin Budding. He invented the lawn-mower in 1830.

When it comes to medicine; in 1796, Edward Jenner found a vaccine for smallpox. In the 1860s Joseph Lister developed antiseptics, which saved probably hundreds of millions of lives. In 1928 Alexander Fleming discovered penicillin, which set the foundation of modern antibiotics.

Finally, whenever you eat a sandwich remember the Earl of Sandwich, who wanted food that was easy to eat while gambling. Whenever you drink milk or eat a dairy product, remember that Leighton Colvin patented the first practical milking machine. Where would the world be without canned food? Thank Peter Durand who made a can out of soft steel and coated it with tin. Charles Strite invented the pop-up toaster in 1919.

This is being written in Niagara Falls, NY, USA. At the time of this writing, the tallest building in nearby Buffalo is London's HSBC. The largest employer in Niagara County is the company that is descended from Harrison Radiator. All around are Kentucky Fried Chicken stores as well as Taco Bells, Pizza Huts and, Dunkin Donuts, all British-owned. Noco gasoline stations are owned by British Petroleum. The local electric utility, Niagara Mohawk, is owned by Britain's National Grid.

This does not even begin to account for those of English extraction that have made contributions in other countries. For example, probably about 80% of the blood in the history of the American presidency has been English.

Not too bad for the old country.

INTRODUCTION

The purpose of this book is to contribute new ideas toward helping the world along into the future. I also wanted to introduce the idea of thinking of and classifying new ideas by pattern rather than by subject. There were several ways that I could have organized these ideas but I wished to do so by the pattern of the idea rather than by 'ideas for cars', 'ideas for houses' etc. As far as I know, new ideas have never been broken down by pattern and I believe that this will help us to see potential solutions that have been missed thus far.

The main theme of the ideas themselves is not so much the creation of new technology but getting more out of the technology that we already have. Just because something has already been thought of does not mean that it has been fully developed. Even though a technology may have been around for quite some time, it still does not necessarily mean that we are getting all the possible benefit from it.

Ideas are like an island on a map. The island is what we have already thought of. The area off the island but still on the map is the benefits that we could have with existing technology but do not have because we have not yet noticed them. Finally, the area off the map altogether is the area that is not accessible now but will be accessible when we have the technology.

These ideas are in various stages of development. Some are raw ideas and some are ideas that would not be patentable. No patent searches have been done by me in regard to any of these ideas, I will leave that up to you.

Maybe you can see some things in some of these ideas that I have not seen. Possibly you can take some of the concepts and apply to other things. Keep in mind that many ideas are happened across accidentally while searching for something else. If nothing more, I hope that this book will be an example of thinking outside the box and will demonstrate how many ideas are lying around all around us just waiting to be found.

1

FURTHER ADAPTATIONS

All around you wherever you go is so much waiting to be discovered. There are millions of potential ways to improve the world around us that we have not noticed yet. We have not even come close to getting the most out of the ideas that we already have. Many existing ideas could be taken much further or have many more manifestations.

Remember that your mind is inside a box. The box is made of habits, culture, grooved-in thinking and respect for the world as it is. If you could dispose of this box, you would see all manner of wonderful possibilities. A child is free of the box but a child does not have the knowledge and experience to make use of this freedom. By the time we have the knowledge and experience to come up with new ideas, we find our minds inside the box.

An ideal example of a further adaptation is the zero. Roman numerals could express a numerical value but calculations could not be performed. Someone, most likely in India, came up with the idea of a zero and calculations with numbers became possible.

IDEA #1; FRUIT WETNAP: It would be far better from a dietary point of view for someone looking to buy a snack in a store to buy a piece of fruit rather than a candy bar. My impression however, is that most fruit buyers buy it to take home.

The problem appears to be that fruit, other than that in which the skin is peeled prior to eating such as oranges and bananas, should be washed before eating. A candy bar, by comparison, has no such need for washing. Everyone knows that it is much better to eat a peach, plum or, apple than a candy bar but that would mean finding some water to wash it with or taking the trouble to ask that it be washed. Thus, the candy bar usually wins out.

The solution is a fruit wetnap, a small paper towel packed in water enclosed by a package similar to the wetnaps already available using alcohol. These should be available wherever fresh fruit is sold.

IDEA #2; SETTING TEMPERATURE AND PRESSURE FOR FAUCET: We carefully set the temperatures of our houses and ovens. We set the time for our microwave ovens. We set the temperature of the heating and air conditioning in our cars to just where we want it. So why, in this time when water is becoming increasingly scarce, do we waste so much of it while messing around with the tap trying to get just the right water temperature? This does not appear very Twenty-first Century to me and is certainly a relic of less efficient days.

It should be easy to set both the temperature and pressure of the water and to have a standardized system of doing so. Everyone would know the temperatures and pressures that water would need to be for various tasks. They would need only to set a temperature dial and a pressure dial and press an on switch. The faucet would only give out the water at the desired temperature and would stop building pressure when the desired pressure is reached. The technology to do this is extremely simple and would save millions upon millions of gallons of precious water.

If you just wanted some water and did not care what temperature it was at, there could be an override function for the temperature control.

IDEA #3; INFLATABLE COAT: In any cold-weather coat, what actually does the insulating is the air within the coat. Snow and down are both good insulators because they hold a quantity of air. So why has no one yet introduced a coat with air pockets inside that can be inflated and deflated at will by a small pump to create maximum comfort?

IDEA #4; CAR CAMERA: Any photographer will tell you that a good photograph often comes about unexpectedly. However, America is a nation of drivers who spend much more time driving than walking to wherever we are going. Suppose we are driving and see a Kodak moment coming to pass? We have to stop and find a place to park and then take a photo. This is highly awkward and inconvenient, to say the least. Suppose a photographer sees a crime or other incident in progress while driving but is hesitant to expose themselves to gunfire or revenge by stopping and snapping a few photos?

Why not just recognize that driving has largely replaced walking and mount a weather-protected camera on the hood or roof of the car. The camera would be operated electrically with a control box within easy reach of the driver. It is unlikely that such a camera would take a close-up photo so it could be pre-focused on a distance. If volunteer firefighters can place a flashing light on top of their vehicles, then why cannot photographers put a camera there in the same way? Let's face the fact the majority of sightseeing is now done from cars.

IDEA #5; SET CASSETTE PLAYER REWINDING OR FAST FORWARD-ING: The one thing that I never cared for about cassettes is the difficulty in rewinding or fast forwarding to a desired point. At the time of this writing, the cassette has not yet gone the way of the eight-track. There should be a numeric system listed on each cassette so that any desired point could be sought immediately.

A tape may, for example, be scaled from zero at the beginning of the tape to one hundred at the end. A song may be listed on the cassette as beginning at 37. The user would be able to insert the tape into the player, enter in the number "37", press the seek button and the player would find the beginning of that song, whether by rewinding or fast forwarding. All that is necessary is a head that can read the scale without actually playing the tape, probably by placing the scale on a separate "scale track" on the cassette.

I still believe that the humble old cassette tape has quite a bit of life left in it because of it's versatility. Why not join cassettes with computers? Suppose someone has a particular interest, such as soccer or world news. Suppose that their commute to work or school is twenty-five minutes every morning.

Suppose there was a device that held a cassette tape and was connected to the internet. The user could choose their interest and enter in their commute time. This would download the latest news or items of interest in the chosen field and trim it down to the indicated commute time for listening during the morning drive. Upon arriving home, the cassette would be placed back in the device and the next download done for the next twenty-five minute drive.

IDEA #6; HOME AND BUSINESS SHOPPING CART: While doing business to business sales work, I could not help noticing a number of shopping carts that had found their way into businesses to be used for moving things around. Any identifying features from the store that the cart was "borrowed" from were usually removed. In some cases the basket of the cart had been removed and, the base had been modified for some special task.

The fact is that the conventional shopping cart is an extremely useful piece of equipment whether in standard or modified form. However, I have yet to hear of anywhere that anyone other than a store can buy a shopping cart.

IDEA #7; NR-NOT RELATED: Our language has a lot of room for improvement in making communications more efficient. Abbreviations are often used for the sake of brevity and efficiency. For example, we use "sic" for "spelling incorrect" when something is quoted and a word is spelled incorrectly in the original document. Such as "Bill wrote 'He wend (sic) to the store'".

There are a lot of listings of names in which two or more people have the same last name. If there is no family relation between such persons, the trouble is usually taken to explain by various choices of words that such persons are not related. Why not simplify this and make our communications more efficient. After all our language is continually changing. From now on if any list of persons contains one or more with the same surname but with no family relations existing, let's just use (nr) in the same way that (sic) has long been used.

IDEA #8; THE LUNAR MERIDIAN: How should we describe a location on the moon, such as the site of a spacecraft landing? Why not just replicate our system of latitude and longitude on the surface of the moon? It is a system that we are already accustomed to and so this would be the most logical approach.

Latitude on the moon is simple enough. The moon does not actually rotate in the way that the earth does over twenty-four hours. However, the same side of the moon always faces the earth so that the moon does in effect rotate once as it goes around the earth once in about twenty-nine days. So, the moon can be said to have a geographic north and south pole marking the axis around which the moon rotates. The Lunar North Pole is the topmost point on the moon as seen from earth and the Lunar South Pole is the bottommost point. Each pole will be marked as 90 degrees and the Lunar Equator will be marked as 0 degrees just as on earth.

In order to measure Lunar Longitude, however, we need a starting point because such an obvious starting point does not occur naturally. On earth, the corresponding starting point would be the Prime Meridian or, 0 degrees longitude.

What we need is a "Lunar Meridian" corresponding to the Prime Meridian on earth. This will act as a starting point for the measurement of longitude on the moon. Since the same side of the moon always faces earth, my proposal is to use the right edge of the moon as seen from earth as the Lunar Meridian.

The moon orbits the earth going eastward, the same direction that the earth rotates. That means that the new moon begins to wax into a crescent moon from the right side of the moon as seen from earth. The previous moon concludes on the left side of the moon. If the moon orbited the earth going westward, the new moon would start from the left.

For that reason, I believe that the line representing the right edge of the moon as seen from earth and running from the Lunar North Pole to the Lunar South Pole should be named as the Lunar Meridian, the starting point for the lines of longitude on the moon. The visible side of the moon would thus be the western hemisphere and the non-visible side would be the eastern hemisphere. Or, we could use + and—instead and refer to the side of the moon visible from earth as the +, or plus, side.

IDEA #9; ALGEBRAIC SYMBOLS IN DATES: This is another way in which to increase efficiency of expression. In algebra, symbols such as x are used to express variables or unknowns. Why not use this system for dates also?

There are no zeros in days in a month or, months in a year so 0 could mean "any" when in a date. 5/0/02 would mean "any day in May 2002" or simply, May 2002. 0/1/xx would mean "the first of the month (in any month in any year)". It is necessary to use x instead of 0 for the year because 00 was a year (2000). Meanwhile, x would stand for unknown, as opposed to any. 5/x/00 means we know that something occurred in May 2000 but we do not know the exact day. 196x would be another way of saying "The Sixties".

IDEA #10; ARTIFICIAL FOODS: Do you know what the big news is every day? About forty thousand people starve to death every day, if not more than that. Except that you do not see it in the headlines.

On September 11, 2001, it could have read: Forty Thousand Starve to death. And oh yes, in other news three thousand were killed in a terrorist attack. The forty thousand that starve to death daily is really the elephant in the room that people rarely discuss. If it happened in one place, it would certainly make the news but it is dispersed across the world. Another thing that few people really realize is that the way we are going, the edible fish in the oceans will be gone in the foreseeable future.

We have got to do something about this. If we can make artificial rubber, which is composed mainly of carbon and hydrogen, why can we not make artifi-cial food? There is an abundance of organic material in the world, we could separate out that which is inedible and mass produce loaves or cakes of nutritious artificial food that need not be based on anything in nature.

IDEA #11; ROAD MAP SHOWING HILLS BY COLOR: Hills are obvious landmarks when trying to find one's way around an unfamiliar city. On the other hand, it can be tricky for a plains-dweller to find his way around a hilly area. The fact is that hills and elevations are very important in navigation.

However, I have yet to see a city road map showing elevations with any significant clarity. When is there going to be a road map showing varying elevations clearly to make navigation easier? It should be easy to do using background colors.

IDEA #12; IDENTIFICATION BY FOOTSTEPS: Fingerprints have long been used for identification. In recent years, methods of identification such as hand geometry and cornea patterns have been developed. However, another very individual trait seems to have been overlooked as a method of identification. The way a person walks is as individualistic as their fingerprints and can be discerned even with different shoe types and change of bodyweight.

A device called the "hodometer" was invented in 1970 to count and map traffic flow in a building. It consisted of pressure-sensitive pads connected to counters. Today advances in materials science makes all kinds of advances possible.

A device for identifying people by their walking pattern would probably consist of a mat large enough for several steps and placed where normal steps would be taken. It could possibly be concealed. The device could also be made to conceal a camera and to count the flow of foot traffic.

The device would most likely have some type of covering that would lessen the importance of shoe type. It would be pressure sensitive with a memory (possibly a computer). It would operate by measuring, analyzing and, recording the pressure patterns when a person walked across the floor. Most likely this would be done electrically, by having a top and bottom to the device that would produce varying degrees of electrical contact when pressed together by a footstep. The device could very well consist of electrical cells resembling those in microphones covered by a rubberized top layer minimizing the importance of shoe type.

IDEA #13; IMMIGRANTS' DAY: Today there is more immigrants than ever before, that is people who live permanently in a country other than their birth, immigrants have traditionally been looked down upon by the native-born people.

I believe that this attitude is completely unjustified.

The presence of large numbers of immigrants is a sign of prosperity. There is something there that makes people cross the world for.

Many of today's immigrants are jet setters. They might have connections to four or five countries and speak that many languages. An immigrant has a three-dimensional view of the world that a person that has only ever lived in one country cannot match. It is often immigrants that know and appreciate the values of the host country the best.

The words on the Statue of Liberty refer to immigrants as "wretched refuse". I believe that this is demeaning to anyone that has come to live in the U.S. and should be removed. Remember that these words are from the days not long after the abolition of slavery and while American Indians were still being massacred by the U.S. Army.

We should also discontinue the U.S. government habit of referring to non-naturalized immigrants as "aliens". In these days when the United States and it's way of life is not exactly popular, we do not need to offend any more people. If you have any photos of your ancestors that landed in America, do they look like aliens or 'wretched refuse'?

Also I believe that the melting pot idea in the U.S. is dead, if it ever was a reality. While working as a telemarketer, I was amazed at the number of American businesses that do not even answer the phone in English or have answering machines recorded in another language. America will be far better off letting immigrants live their own way after they have taken an oath of allegiance to the U.S. constitution.

In this new international world of immigrants, we should have a term for those people that have only ever lived in one country. I do not mean visits, but permanent habitation. Let's call such an individual a "uninational" or "uni" for short.

We should also have a day out of the year to celebrate all of those among us that were born in another country. We will naturally call it "Immigrants' Day".

IDEA #14; PERMILLE: We have long used the percent system to express a portion or a fraction. 86% means 86 out of every 100. The trouble is that when the percent system came into use, the world was much less complex than today. It seems that now we can measure just about everything with more precision than we could years ago. Things do not fit neatly into our old scale of 1 to 100. We often see figures like 26.1%, 33.4% and, 76.9%.

Why don't we replace the old percent system with the new permille system? "Mille" means thousand in the same way that cent means hundred. We would be measuring on a scale of a thousand rather than a hundred. Until we have a symbol for permille, we could just use pm. 26.1% would be 261 permille. 33.4% would be 334pm. 76.9% would be 769pm. This would be a more accurate system and would recognize the advance of precision in measurement.

IDEA #15; AUTOMATIC WINDSHIELD WIPERS: This one is way overdue. Cars are doing all kinds of things automatically nowadays, such as turning headlights on and off. When are they going to be sensing moisture on the windshield and activating the wipers to the appropriate intensity? Sensors at the bottom of the windshield could detect water. Another sensor in the motor driving the wipers could sense if the wiper blades are worn and adjust accordingly. Since wipers are a very important safety feature on a vehicle, the automatic system would probably have a manual override.

IDEA #16; ARAMAIC LANGUAGE TAPES: Language tapes can be bought in large bookstores to study all of the major languages of the world. I have seen Latin study tapes. Latin is no longer a living language but still has it's devoted students.

One that I have yet to see however is Aramaic. It has not been a living language in a long time, approaching two thousand years. The word "amen" is believed to be the last remnants of the language.

What is significant about Aramaic is that it is the language that Jesus spoke. I am sure that many Christians would be interested in at least hearing the language of Jesus and possibly a recreation of His Words as recorded in the New Testament in the original language.

Aramaic was the language of the Arameans, who lived around what is now known as Syria. They were a group of people who never had a large, contiguous empire but who were so active in trading that their language became widespread in it's use.

IDEA #17; MULTI-BAND RADAR: Radar is an acronym for "radio detection and ranging". It works by sending out a high-frequency radio wave, using a directional antenna. When the signal encounters a metal object, it sets up a faint alternating current in the object just as a radio station does in your car radio antenna.

The frequency of this alternating current is the same as the original signal and so this current sends out electromagnetic waves of it's own at the original frequency. In effect, the signal emanates from the original antenna and "bounces" off the metal object.

The original antenna also receives radio waves at the same frequency as it transmits. The radar station has both a transmitter and receiver. Thus, the faint signal from the metal object is received back at the radar station. The speed of electromagnetic waves is known, 186,282 miles per second or 300,000,000 meters per second. Therefore we can determine the distance to the metal object by incorporating a timer into the system and measuring the time between the transmitted pulse and the received signal.

Since a radar antenna usually rotates in a circle, it can easily be determined the direction of the metal object from the radar station. All the information can be sent to a cathode ray tube and the object will appear as a blip on a screen. Some frequencies are reflected from water droplets and so can be used in weather forecasting.

My idea for improving this system is based on the fact that an alternating current, creating a radio wave, goes out into the conductor and then returns to the starting point in one wavelength. The voltage that produces the current moves through the conductor at the same speed as an electromagnetic wave travelling through space. This means that the most efficient antenna for broadcasting a radio wave is one that is half the wavelength. (The length of an electromagnetic wave is the speed of the wave divided by the frequency).

Why don't we have the radar station send out waves of several frequencies, instead of just one frequency. A metal object, such as an aircraft, will reflect all of the waves back to the radar station. However, since the most efficient antenna is one of half the wavelength, it will reflect some frequencies a little more strongly than others. If we use a wide enough spread of frequencies, we could determine at which frequency we received back the peak signal strength. Since wavelength can easily be calculated from frequency, the only thing necessary is to divide the wavelength in half to determine the length of the object.

If we set up several receiving antennas a short distance apart, we could determine the geometric plane in which the waves were radiated from the aircraft (An antenna naturally sends out waves in the same 3-dimensional plane in which it is positioned). This would tell us immediately it's angular attitude and direction of flight. This is a lot more information to be gained than just a blip on a radar screen. There are multi-band radar now but I have yet to hear of one using what I will call the "Half-Wavelength Principle".

IDEA #18; MICROSCOPIC PHOTONEGATIVE MARKER AND CUT-TER: If you have ever taken a photography class, you know that there are many interesting ways to manipulate a photograph in the darkroom.

However, one that I have yet to see is manipulation of the negative itself. We could have a specially adapted microscope and tools to perform "surgery" on a photographic negative. Different negatives could be spliced together. Markings could be made or altered on the negative. Dyes could be added or removed. This is yet another way for conventional photography to compete against digital cameras.

IDEA #19; DRIVER WARNING LIGHTS FOR CAR LIGHTS: Do you want to know what cars need a warning light for? When a light burns out, that's what. It is difficult for many drivers to know when a headlight goes and very difficult to know when a taillight burns out. Just put a light on the dashboard and have it light up when the headlights, taillights and, brake lights should be on but the circuit is broken.

IDEA #20; USE OF ELECTRICAL CAPACITANCE FOR MEASURE-MENT: I am convinced that this is an idea with tremendous far-reaching potential. A capacitor is simply two parallel metal walls in close proximity. A capacitor stores an electrical charge and then releases the charge when it's capacity is reached. Capacitors are in wide use in electronic devices. The basic unit of capacitance is the 'farad' and is named for Michael Faraday. Capacitors usually consist of strips of metal foil separated by an insulator of some kind, rolled up and placed in a protective casing. There are also so-called "electrolytic capacitors" which use a liquid between the walls. Radios tune with a "variable capacitor" in which a movable set of metal plates is set up with a fixed set of plates.

My idea is based on the fact that when anything is placed between the walls of a capacitor, it changes it's capacitance. We could set up a large-scale capacitor having two metal walls insulated from each other and connect an alternating current. With the extreme sensitivity of electric meters, this setup could have all kinds of uses. It could tell how much garbage is in a dumpster. It could check air ducts for debris. It could warn of intruders in a room. It could identify people walking through a hallway. It could ensure that the correct item was in a box. It's uses would be almost limitless.

IDEA #21; THE SPECTRAL HORIZON: In recent years there has been a tremendous improvement in graphics of all descriptions. This has made easier what I will call "The Spectral Horizon". I believe that we could and should be making much more use of color-coding. The terror attack warnings being used today could be only the beginning.

Many things are already color-coded. Traffic lights are the first to come to mind. Blue stands for boys, while pink stands for girls. Electrical resistors make use of the visual spectrum to indicate the number of ohms in the resistance. The colors of the spectrum, from lowest to highest frequency, are; red, orange, yellow, green, blue and, violet. In electrical resistors, the colors white, black and, brown are also used. Colors are used for the state of alert for terrorist attacks.

There should be an international color code system. To start with, the four cardinal directions; east, west, north and, south should be assigned a color. Opposites such as up, down, past and, future should have their own colors. In the case of opposites; one could be indicated by a hot color-red, orange or, yellow; the other by a cold color-green, blue or, violet. The days of the week and the months of the year should be color-coded.

Color is universal, understood by people of all languages, and has always been something of great sensitivity to humans. In the age of graphics, color is too useful to not be used to the maximum. We are now on the Spectral Horizon, the future should make use of color-coding much more than it does now.

IDEA #22; IMPACT CLOTHING FOR FALLING: Have you studied people falling? In movies, sports, etc. Whenever people fall on a flat surface, the same points on the body almost always land first. The impact is almost always taken by the knees, shins, hips, lower back, the elbows and, the area around the shoulders.

I was wondering why no one has ever made "Impact Clothing" to protect people engaged in some occupation or activity in which a fall is likely. Falls could be closely studied and the clothes padded accordingly. This would be certain to prevent a lot of injuries in certain occupations.

IDEA #23; BODY STETHOSCOPE TO LISTEN TO DIGESTION: In study of the digestion process and treatment of the disorders thereof, one factor that has been virtually ignored is sound. The digestion process is audible. If sound is a useful factor in analyzing the performance of an automobile engine, maybe it could be useful in the digestion of foods and liquids.

A stethoscope is used to help a doctor determine heart condition. Why not modify a stethoscope by combining it with an abdominal girdle to listen to the digestion process and see how much this new tool can help us.

IDEA #24; ANGULAR MEASUREMENT SCALE: When someone looks through a magnifying device such as a telescope or binoculars, distant objects are seen as if they were closer. Any such scope could be made much more useful by adding an unobtrusive scale of angular measurement. That way, an angular measurement could be made of distant objects or of the angular distance between objects.

There is at present no device except for surveyors' scopes to visually measure angular distances at a distance. If the scope only has one possible power of magnification, the scale would most likely be inscribed on a glass plate that would be inserted before the objective lens. If the scope had a range of possible magnification powers, a method of adjusting the degree scale when the magnification is changed would be needed, this would most likely be accomplished by the use of two inscribed glass plates in which one would move when the magnification of the scope was changed.

Such a scale would make a magnifying device much more useful. By using the angular measurement scale, it could be determined how far apart objects were in the view if the distance to the objects from the observer was known. Or, the distance from the observer to an object in the scope could be determined if a distant object could be found whose size was known. At any rate, the distances between objects or the sizes of objects could be quantitatively compared at a distance.

IDEA #25; PHOTO PAPER TEXTURED TO RESEMBLE WATERCOLOR: Many times a photograph is taken not for accuracy but for beauty. This is especially true in landscape photography. A watercolor done very well resembles a photograph. Why not manufacture photographic paper with a coarse texture so that the photograph will resemble a watercolor? I have some photographs that would look outstanding as watercolors. The graininess of a photograph when enlarged would not matter as much if it was intended to resemble a watercolor.

IDEA #26; SMART TRAFFIC LIGHTS: Nowadays there seems to be so much talk about so-called "smart" devices. From smart bombs to smart cards to smart appliances the days of dumb things seem to be numbered. So why then is there still a device that we encounter every day and does so much to conduct the flow of life for which "dumb" would be almost a complement? I am referring to the traffic light.

A traffic light is a supreme example of inefficiency, especially in our rushed and time-starved era. We simply must do something about this. How many times have you seen a traffic light at an intersection with a line-up waiting for the red light to change but no cars going through the green light side? Considering the nation as a whole, this is a phenomenal waste of time and fuel as well as a decrease in the quality of life and an increase in pollutants as well as wear and tear on the vehicles. When a car stops, it loses the momentum it has built up and must waste fuel in idling and then accelerating.

There are a number of ways to improve traffic light technology. Smart traffic lights could send out a directional super-sonic pulse and listen for it's echo to sense the presence of waiting cars. Possibly at night, when traffic patterns are not as predictable as in the daytime, it could sense red light reflected back from windshields of waiting cars. Of course weather conditions, which may include the light swaying in the wind, would have to be considered.

It would be great if we could just have an underpass-overpass system at every intersection so traffic lights would not be necessary. Until then, let's make the lights as efficient as possible. At this point, just about any change would be an improvement. If you could travel into the future and look back at our time, one thing that would stand out as utterly archaic would be lines of vehicles waiting at traffic lights.

IDEA #27; MEASURING DISTANCE WITH FLASHLIGHT ATTACHMENT: During darkness, a flashlight could be used to measure distances in surveying. It could be done with a specially designed flashlight or an attachment to fit onto a flashlight.

Suppose we had an aperture dial to fit over a flashlight so that the beam could be varied in size. It could include a calibrated dial that would indicate the ratio of the light disc diameter on a distant wall to the distance from the flashlight to the wall. All we would have to know is the width of the light disk to calculate the distance to the wall. To do this we could simply place to chalk marks or other markers a known distance apart, then adjust the light disk until it just fit over the markers.

It may be easier to use a slit that is variable in length instead of an aperture. Instead of varying the length of the slit, shadow-casting markers with an attached scale could be used to cover the width of the markers. The scale, of course, would indicate the ratio of the light bar (or disc) on the distant wall to the distance of the flashlight from the wall.

IDEA #28; LATERAL VIEW AUTOMOTIVE MIRROR: When they decided to put mirrors on cars, to see behind and on the sides, there is one place they forgot. I have yet to see a car with a mirror where the hood ornament would be. The mirror would be angled to permit the driver to see to the left in countries that drive on the right side of the road and vice-versa. This mirror would not always be needed so it could be retractable.

When the snow is piled up at the side of the road and a driver is stopped at a stop sign waiting to turn onto the perpendicular road, he/she has a difficult time seeing oncoming traffic. Many drivers in such a situation try to inch forward but this creates a dangerous situation. In some places, Europe in particular, the view when exiting a narrow lane may be blocked by buildings. The problem would be solved by a look in what I will call the "Lateral View Mirror".

IDEA #29; PASSENGER COMPARTMENT REAR AIR VENT: It can be very pleasant to drive in the nice weather with the window down. However, auto manufacturers do not seem to consider this. They assume that the air conditioning will be used. Driving with the window down creates an annoying draft, especially if there are back seat passengers.

Why not put a vent at the top of the middle of the back seat in the car? That way, the air would move in a smooth flow from the window into the car and out the rear vent. As it is now, air rushes into the car through the open window and has nowhere to go and so produces a swirling draft.

IDEA #30; ELECTRONIC MENUS IN RESTAURANTS: This is another one that is really overdue. Why cannot restaurant customers click on a menu choice and see a complete photo and description of the choice. There would be a number of ways to implement this, the screens do not necessarily have to be on the tables. Traditional menus could still be available for those disinclined toward computers. The computers could be set up in a restaurant intranet. The waitresses could be called via the computer.

IDEA #31; DOUBLE-CHAMBERED MOP BUCKET: Have you ever noticed that when you mop a floor with a single chamber mop bucket that you put quite a bit of the dirt back on the floor? Why not divide the bucket into two chambers, a smaller chamber under where the mop is wrung out and, the larger chamber to hold the water. The smaller chamber would hold the dirty water wrung from the mop instead of letting it mix with the clean water in the same chamber. Do some experiments to find out what percentage of the water in a mop bucket ends up being actually wrung from the mop during a typical mopping (remember that some water will remain on the floor). Then it will be known what percentage of the bucket should be included in the smaller chamber for the dirty water.

IDEA #32; LEATHER COAT MADE ESPECIALLY TO MATCH HAIR COLOR: I think that this would be a neat fashion idea. Since colors have a history of being so important in clothes matching, I am surprised that this is not available yet. It is relatively easy to make leather in different colors and this idea does not have to be limited to leather coats.

IDEA #33; MOTION CLOCK: Traditionally, clocks have told the time by the use of a big hand and a small hand against a backdrop of numbers from one to twelve. The small hand goes around once every twelve hours and the big hand goes around once every hour.

This is all well and good but who says we have to be limited to this particular clock design? Why not introduce some artistry into the process and come up with some other clock designs? We could have an entire story told over the course of a day and containing sub-plots taking one hour each. One could tell the exact time by how things were positioned in the story and sub-plots. With today's technology making this easy to do, this could be a whole new art form.

IDEA #34; SMART LAWN SPRINKLER: The trouble with most lawn sprinklers is that they waste water. Today, when water is becoming more and more precious, we could really use a sprinkler that does not waste water.

What is needed is a sprinkler that is adjustable to exactly fit the dimensions of the lawn in it's discharge. If the sprinkler is placed in the center of a yard, this would require it to throw water further at the angles toward the corners of the yard than to the midpoints of each side of the yard. Once the dimensions of the yard and the position of the sprinkler are programmed into it, water would automatically be thrown the correct distance in each direction, greatly reducing water waste.

IDEA #35; DIRECTIONAL SIREN FOR EMERGENCY VEHICLES: Emergency vehicles often give people trouble in determining their direction of travel. One can usually tell when an emergency vehicle passes due to the drop in sound caused by the doppler effect. But still, emergency vehicles typically cause confusion as to where they are coming from and where they are going. For the most part, sound from the siren is only needed in front of the vehicle and to a lesser extent from the sides.

Why not set the sirens on a vehicle so that there is two sources of sound for each frequency. The sources will be placed so that they interfere constructively and destructively according to the principles of wave interference to "beam" the sound directly ahead, although a certain amount of sound will also slip out to the sides.

IDEA #36; SEASONAL CALENDAR: What really causes the difference in changes in our way of life throughout the year is not the months but the seasons. We do things by seasons much more than by months or year. I have wondered why we have always used months. It was almost certainly for the purpose of planting since the moon provides a convenient beacon by orbiting the earth about once a month.

In the twenty-first century, only a few percent of the people in the advanced countries work in agriculture. In any case, it has probably been a long time since anyone has relied on the moon to plant seeds at the correct time. Why not just use four seasons instead of twelve months? This would be especially fitting since we have a leap year once every four years and the extra day could be given to a successive season every four years.

Another issue is that seasons begin on the twenty-first or twenty-second of the month. I wonder why someone did not coordinate the system so that seasons begin on the first of the month. Also, I cannot figure out why New Year's Day is January first. It was because of Julius Caesar but would it not have been better for the New Year to begin at one of the solstices?

IDEA #37; PLUS AND MINUS TIME: Another thing that seems archaic in the new millennium is the use of A.M. and P.M. when telling time. It would look much more modern and digital to use "-" for midnight to noon and "+" for noon to midnight, just as an integer can be negative or positive. It would also be less likely to cause confusion when being translated from one language to another.

IDEA #38; AUTOMATIC CHECKOUT SCANNERS: There is much more room for automation and convenience in supermarkets and department stores. If we can have alarms that tell if anything is leaving the store, why can't we have entirely automatic checkouts? All that would be necessary is to encode the information on bar codes about each product onto magnetic strips and then sense it magnetically. Just push the shopping cart through a structure with sensors and it will itemize everything in the cart and ring it up. There are already checkouts where the customer rings up the purchase but this would take it one step further.

IDEA #39; DECIDUOUS TREE LIGHTS: If we light up evergreen trees at Christmas with strings of lights, why not light up deciduous trees the rest of the year? Deciduous trees are the trees with leaves such as maple, oak, elm, sycamore, etc. With their broad leaves, I think that lighting kits for deciduous trees will produce spectacular results. Lights, connected by electric wires as in Christmas lights, could be fastened to the trunk and branches of the tree so as to shine on clusters of leaves. This light would illuminate a yard without the harsh glare of unshielded lights.

IDEA #40; THE FREEDOM BOX: What surprises me about the nations that give "freedom" to it's citizens is how little discussion there is after grade school about what exactly freedom really means. The theory is that human beings perform better when they are free and there seems little doubt about the truth of that theory.

However, that still does not tell us what exactly freedom is and how we can improve or perfect this concept of freedom. Freedom cannot be to do whatever you want such as seeing someone stopped at a traffic light driving a car that you like, simply shooting the person and taking the car (although that does happen). That would be the law of the jungle. What men such as the founders of the USA intended was a civilized freedom.

We do of course have laws to guide our behavior. In the U.S., there is also a written constitution. However every law or body of laws has two components, the letter and the spirit. Jesus, for example, placed much emphasis on the spirit of the law and not just the letter.

Everyone knows that the freedom practiced in western democracies is far from perfect. Freedom, after all, is a somewhat radical idea if we consider the whole of human history. Those who dislike the United States or other democracies often point out how shallow the freedom practiced there really is and the inability of people to really handle such principles.

It does seem as if there is an "official society" and an "unofficial society". The official society sets down the law giving freedom to it's citizens. However, there emerges an "unofficial society" that runs counter to the official society. The unofficial society is unwritten and often unspoken and does much to take away the freedoms given by the official society because people cannot really handle it.

The unofficial society does not consist of officialdom. It is made up of the "crowd" and the "community". It sometimes seems as if a country like America has just turned the pyramid upside down so that instead of being ruled by a king or dictator, we are ruled unofficially to a large extent by the people around us.

While many may think that "community standards", "going with the flow" and, "toeing the line" is for the common good, it cannot be denied that this "unofficial society" is counter to the constitutional freedoms that people are supposed to have and does give credence to the argument that people as a whole have difficulty really handling freedom.

Regardless of the individual rights that citizens of free democracies are guaranteed, in many circumstances personal privacy is non-existent and bored people or those who feel threatened by those unlike them, and offended by those who do not want to be like them, practically make a career of getting into others' business.

There are those who think that this is for the better but what about the eroding of the hard-won principles of free societies? Are we really free if we have to have the "right" friends, have the opinions that we are "supposed" to have and, think the way we are "expected" to think? In this respect, the unofficial society, free societies are in fact mirror images of those that are not free.

To help perfect the idea of personal freedom, I would like to introduce the idea of dimensions. Civilized freedom, as opposed to the law of the jungle, can be compared to a box. The box is made up of the laws and constitutions that define our freedom. The law of the jungle has no box, it is represented by an open space, one may do whatever one wishes but anyone else may do whatever they wish to you.

Usually in coming up with new ideas, we seek to think outside the box. This is the one exception. To have a free but civilized society, we come inside the box of the law and principles.

The law of the jungle has no legal structure and is defined as 'freedom to'. Civilized freedom has a legal structure (our box) and is defined as 'freedom from'. In civilized freedom, you are not free to steal someone else's possessions but no one is allowed to steal your possessions. The box, as opposed to an open space, represents 'freedom from' as opposed to 'freedom to', which is the law of the jungle.

The box of freedom has three dimensions just like any other box. Thus there are three possible dimensions to human freedom in a free and civilized society. Zero dimensions would be a confined existence like being in prison. One-dimensional freedom would not be prison but would be a very ordered existence akin to life in a dictatorship. Two-dimensional freedom would be like our society, you are given the legal rights to go where you want, do what you want, own what you want and, be what you want as long as you do not infringe on the rights of others to do the same.

So if we are already free after two dimensions, what is the third dimension of the box? I define the third dimension of the freedom box as "freedom from unreasonable interference". If you have an unofficial society around you that forces you to "be accepted", "get into the clique", "be what people want you to be", "be like and think like everybody else" or, "go with the flow" and that these pressures are not in accordance with the officially given freedom, then you are living in a condition of "flat" or two-dimensional freedom. You are still freer than one in a dictatorship but you do not have full three-dimensional freedom.

I believe that the founders of the United States of America intended for U.S. citizens to have full three-dimensional freedom. The interference directed at a person or persons of major age in a way that is not written in the law or constitution is limiting freedom from the full three dimensions to an inferior two-dimensional freedom that is merely one dimension ahead of a dictatorship.

The reason that this happens is that so many people cannot really handle freedom. It does take some special people to handle these principles. It is not within the spirit of the U.S. constitution to set up a community so that everyone has to be monitored by strangers or to put a definition on someone in the community or to set it up so that someone of a minority group is discouraged from celebrating themselves. It is extremely un-American to try to put limits on a person because they have interests or a background that you do not have.

In my opinion, the idea of freedom contains some high principles. As with all high principles, the great challenge is to prevent erosion of the principles. Erosion usually begins in the "unofficial society" when people do not really have the spirit of the principles. Principles may start out with both the letter and the spirit, then there is the letter but people have lost the spirit. Then, one day the principles are gone.

The great enemy of all high principles is convenience. Traveling by the high road leads to a better life but is also more difficult and less convenient.

High principles must exist as entities in themselves, not just as a feature of the country. It is not acceptable to erode America's basic principles to make the country a little bit safer from attack. In the long run, we will pay a much higher price for eroding the principles. For us to abandon our principles is what those who hate our principles wish more than anything.

When America was attacked, what were the attackers' objectives? What is just so that they could say; "ha ha, we got you"? I think there was much more to it than that and that they spent years in planning.

First of all, the attack on the WTC was intended for visual impact. If the intention was simply to kill as many Americans as possible, it would have made more sense to hit both towers at the same time so that any evacuation would not be possible. As it was, there was twenty minutes or so between the impacts of the hijacked aircraft. This was planned, I believe, to allow time for hundreds of cameras to let the world see the second impact.

But what did Osama bin Laden really want out of all this aside from inspiring and motivating future warriors?

First of all, he wanted infidels out of the holy land of Islam.

In particular, he wanted to seriously damage the airline industry that has brought about globalization.

He certainly wanted to separate the infidel west from those Moslem nations friendly to the west such as Saudi Arabia, Egypt and, Kuwait.

I believe he wanted to provoke a backlash against those Moslems that would go and live in the west.

However, I believe that most of all, he wanted to shoot down these dangerous principles of freedom and democracy that come from the west. These principles threaten Osama's view of the Islamic world as it should be. I also believe that Osama bin Laden thought he could accomplish this in one carefully planned and thought out move, the attacks of 9/11.

The planes would be hijacked by nationals not of nations traditionally considered as enemies of the west, such as Iraq, Iran and, Libya; but by Saudis, Egyptians and a Kuwaiti. This was calculated to anger Americans against their allies in the Middle East. It would make many people afraid to fly and damage the already fragile airline industry. It would hopefully bring back Americans' anti-foreigner sentiments and separate the U.S. from the Middle East. It would probably also divide America from the U.N. and European allies. It would certainly provoke a backlash against Moslems living in the west and so discourage other Moslems from going there.

In short it would stop globalization, permanently divide the Middle East from the west and let Moslems live the way they always have. I am certain that this is what Osama bin Laden was planning.

But even more than this, he wanted to destroy the western ideas of freedom and democracy. Just as the Mujahedin in Afghanistan stopped the communists,

Al Qaeda would stop democracy and freedom. But instead of nine years of fighting, they would set up the erosion and collapse of freedom in just one day, 9/11.

Osama bin Laden knew from American history that America has a tendency to throw away it's principles when it is angry or feels threatened. He knew about the internment of Japanese-Americans after Pearl Harbor. He knew about the McCarthy witch-hunts of the 1950s. He knew that democracy often works better in theory than in practice.

Not only did Osama want to stop globalization, he wanted America to be a closed and oppressive society rather then a free and open society. He wanted to get rid of these ideas of freedom and democracy.

Ironically, when we lock up prisoners of war and ignore the Geneva Convention, when America ignores the U.N. that it was so instrumental in founding, when we detain people with no charges or customary legal procedures, when we violate long-established freedoms indiscriminately, we are doing exactly what Osama bin Laden wants us to do. I wonder if he put a lot more thought into these attacks than the U.S. government put into the response to the attacks.

Let's have a lot more discussion about what freedom really means and show the world that we really live by these principles in letter and spirit.

IDEA #41; TOWELS WITH OWN DESIGN: We can order many things from T-shirts to lampshades with a design (or possibly a photograph) of our choosing on it. I think that it is about time we could have our own designs put on towels. Wouldn't it be nice to have towels with an image of your house or car? How about gift towels from a company with the corporate name and logo? Towels are very visible and would be an effective advertisement medium.

IDEA #42; BATTERY INDICATING WHEN SPENT: Why is it that we still have to guess whether a battery is spent or still contains power? It is the new millennium now and battery life must still be guessed at or tested. Devices powered by batteries can have a light or other indicator telling the user if the battery is still good. However, I think it best to have an indicator built into each and every battery revealing the amount of life remaining. Simple color codes sampling the internal battery chemistry will do.

IDEA #43; MAP OF ALL POSSIBLE PATHWAYS THROUGH COMPUTER APPLICATION: I think this is an idea that would not only make use of an application easier but would also make sure the user realizes all that the application can do for him/her. It would probably be a flowchart showing all the different pathways that the user can take through the application and would be shown on the application's opening screen or soon thereafter.

IDEA #44; COLOR STANDARDS SCALE: Why do we not yet have a world-wide standard scale in where colors are represented by numbers? We actually do have such a scale and it is used in computer graphics, particularly HTML. We should expand on that to create a general use scale. It would represent all visible colors by a number. The scale would be standard across all industries and would span the entire visible spectrum. We express all other electromagnetic waves in terms of frequency, such as WDCX FM 99.5, why can't we do the same for visible color? It may not be quite the same expressing your favorite color as a number but it would make for much more efficiency of expression, especially in industry.

If we could express colors as numbers, we could make a "world flag". Break a flag down into pixels and express the color of each pixel as a numerical value. Now do this for all of the flags of the world. Take the average value of each pixel and we have our world flag. It is true that some flags are different in shape than others but I do not think that this will cause much distortion for the world flag.

IDEA #45; THE CYBERFORCE: There have been armies and navies since ancient times. The two were considered as separate branches of the military because they each fought their own war against the enemy in their own domain, land or water. In most countries, the navy was independent from the army and vice versa.

Within the last hundred years, the airplane became so important in warfare that most nations have a separate air force to engage the enemy in the air and conduct raids from the air, independently of the army or navy. The air force became the third branch of the service.

The navy does not control all ships, the air force does not handle all aircraft nor do all ground troops belong to the army. That is not how it works. The navy has many aircraft based on navy aircraft carriers. The army may have troop transport ships. The air force may have it's own ground soldiers to guard air force bases.

The navy has special ground troops that it carries aboard ships to make quick raids on the enemy. These navy troops must be tough and resourceful since they cannot count on the backup available to conventional army soldiers. These navy troops were traditionally known as marines.

We can see that it is technology that brings about new branches of the military. When men could build warships, a navy came about that was independent of the army. The same happened with airplanes. Early aircraft were at first considered as just reconnaissance and air raiders for the army. But as warplanes became more and more important, air forces became a full-fledged branch of the military, separate from and equal to the armies and navies.

Since technology is always progressing, what is going to be the next branch of the military? In the opinion of this writer, the next logical branch will be the cyberforce. Cadets will be interested in computers to begin with. Dressed in their uniforms, the cybermen will sit in front of computers and attack the enemy as well as engaging in information warfare and a lot of information gathering. They may try to become pen pals with individuals in enemy countries.

Officers will write programs that will hamper the enemy's efforts and interfere with his communications channels. They will also protect friendly web sites from enemy hackers. Combatants may be individual hackers in neutral countries that sympathize with one side or the other.

IDEA #46; THE MOLECULAR PROGRAM: Why not write a computer program with a list of all of the chemical elements? The user could click on elements and the program would display a list of all the compounds that could be made from the list of elements that the user clicked on and the properties of those compounds.

IDEA #47; NAMING WEEKS: Have you ever noticed how often people express time in weeks? Not as often as in days or months but still, "The first week of April", "The week after Thanksgiving", "The week between Christmas and New Year". The week of the Fourth of July" are common expressions.

Why not name the weeks of the year? I believe that we will find this actually a more efficient way to express time than in months. We probably only measure time in months because the moon readily defined months for us. Today, in our less pastoral times, we almost always mean Monday through Friday when expressing time in weeks. Remember that in the Bible the second measure of time mentioned, after the day, is the week. Face the reality that the lives of most people are defined not by the month but by the week.

IDEA #48; WRITING SOUNDS: Have you ever wondered why words and music can be expressed in writing but other sounds cannot? If you wish to convey the sounds of cars rushing by on the highway, you can make a recording of it or can write "The sound of cars rushing by on the highway". But, there is no way to directly express the sounds as writing in the way that can be done with music by a songwriter. Why has no one yet come up with a way to use the letters of the alphabet and the numbers to express all sounds? It would be necessary to express the three dimensions of the sound, it's loudness, it's duration and, it's frequency. Special symbols like those used in music may be developed and used if necessary.

IDEA #49; SIGN HEATER: If you live in a northern climate, how many times have you seen signs in the winter covered with snow? If we can have heating wires in car windows to eliminate snow and ice, why not have then in signs? Many signs use letters that stand out from the background and often end up with a pile of snow on top of the letters. Simple heating wires, which could be automatic when snow is sensed on the sign, would solve the problem.

IDEA #50; DRAWING COLORS IN BLACK AND WHITE: The trouble with a drawing done with standard black pencils is that you cannot tell what color things are. Some things are obvious like the grass being green, snow is white and the sky is blue, but what about the color of everything else?

I realize that many drawings are artwork and not intended to convey the maximum amount of information. However, I believe that a system of encoding colors onto a black and white drawing will be of great benefit. There are, of course, colored pencils but these tend to be lower in quality and in any case will not reproduce when the drawing is copied on a non-color photocopier.

Let's take the visible spectrum from one end to the other. The visible spectrum consists of red, orange, yellow, green, blue and, violet. White is a mixture of all colors and black is the absence of light and neither require coding. When an artist shades an area, why not arrange a system that can convey the color of the shading? If shading lines are long that might mean the color is of long wavelength, or red. If the shading lines are short, that would mean the wavelength of the color is short, or blue.

The gaps in the shading lines would, of course, be staggered so that it does not appear as if the gaps are a part of the subject. If the color is a pastel or not a primary color, lines could alternate to convey the color mixture. The viewer would obviously need some understanding of the properties of color. Lines would still be close together or darker to convey a dark color and further spaced or lighter to convey a lighter color.

Probably the only reason that a system like this has not yet been implemented is that drawing is a very old form of art dating back to prehistoric times and the physical properties of light and color have only been really understood for a comparatively brief couple of hundred years.

IDEA #51; DUAL GASOLINE INPUT ON VEHICLES: How many times have you pulled into a gas station with some cars already there and had to wait to get to a pump because your gasoline input was on the wrong side for the open pump? Or maybe you were in a hurry so you had to drive around and enter the gas station from the other side.

If cars can have dual exhausts, why not dual gasoline inputs? One input on each side of the car. I am sure that this would not be too difficult and I am sure that the first car to offer this convenience will sell more than it would have otherwise.

IDEA #52; ADJUSTABLE TOASTERS: Why can't you toast a bagel or a bun in a toaster? Because the slots on the toaster are made for slices of bread and are not wide enough, that's why. If the slots were made wide enough to toast the standard bagel, they would be too wide to toast the standard slice of bread with the same efficiency. Why not make a toaster in which one of the sides with the high-resistance nichrome wire that does the toasting is movable to accommodate any bread slice or bun or bagel?

IDEA #53; HAND WASHING AND DRYING PODS: Washing and drying one's hands always go together. Heat from electric dryers is largely wasted when it is directed downward. When is someone going to come up with a pod in which the hands are inserted, washed and then dried? Each pod would be a sink with some way to control the flow of water as well as it's temperature and a way to turn the water off and the dryer on. As an added feature, the dryer could turn off automatically when the hands were removed.

IDEA #54; PLANE SPOTTING: All day we can see the vapor trails of airplanes crossing the sky. Now that we have the internet, why not make a hobby of plane spotting? We know that jet airliners usually travel at 500-600 mph. If we knew where the planes had left from and the destination, it would be relatively easy to identify a distant plane high in the sky. Such an activity would be beneficial in teaching children navigation, geography and mathematical skills.

IDEA #55; COMFORT WITH CITIES: This idea is about an attitude change. It is now the Twenty First Century and one of the major global trends over the past one hundred years is urbanization, the move from the countryside to the cities. Why then are we still not entirely comfortable with cities?

This discomfort is reflected in the names that we tend to give housing developments and urban areas. How many housing developments in the middle of big cities are given names such as "Green Meadows" or "Forest View"? Not to mention the thousands of streets in major urban areas with names or prefixes like "Elm", "Maple", "Sycamore", "Oak", "Pine" or, "Cedar" and suffixes like "-wood" or "-dale".

Where do you suppose a place called "Willowdale" is to be found? Isn't it a bucolic little town? Doesn't it have beautiful willow trees, swaying in the summer breeze? Isn't it filled with the sound of chirping birds? Living by the timeless ways of nature?

Actually, Willowdale is in the middle of bustling, cosmopolitan Toronto.

Why not emphasize our modern urbanization by our choice of names? Where is the corner of Skyscraper Street and Skyline Avenue? How about Downtown Street and Urban Avenue? How can I get to Concrete Street and Asphalt Avenue? What about Cinderblock Boulevard and High Rise Street? I am looking for the City View Housing Development.

IDEA #56; THE BASE LANGUAGE: I got to thinking, numbers and words are actually very similar, they both use characters to represent everything that humans encounter and deal with. Digits can make an infinite number of numbers. Letters can make an unlimited amount of words.

There are, of course, differences between numbers and words. A primary difference is that numbers are universal while words exist in a multitude of languages. Numbers are usually absolute in meaning while words can be much more relative, with shades of meaning. Numbers are more efficient but words but words have more beauty and can convey more meaning. Words are like a painting in the same way that numbers are like a photograph. Words are open to artistry. Numbers are more like cold data.

Despite their differences, we represent things in life by words. We represent the world around us by numbers and the two largely overlap. Those who deal with words and those who deal with numbers may be diametrically opposite cliques but the truth is that mathematics is a language and is actually the world's common denominator of languages.

Increased telecommunications over the last several decades has led to language standardization. Unfortunately, just as letters combine to form words, words combine to form meanings. This is the reason that the words in a document in one language cannot be literally translated one by one into a meaningful document another language. It is the meaning that must be translated. For example, we cannot simply translate the English question "How does it work?" into French because in French one would say "Comment ca marche?" to ask the same question but that literally translates into "How does it walk?" Phrases used by Americans such as one "having a few screws loose" or by Britons such as one being "round the bend" obviously will not carry their meaning if literally translated into another language.

My idea is to take everything that people say to one another and break it all down into numbers. This set of numbers representing the things that humans say to each other will be called the Base Language, in which meanings will be represented as numbers. All meanings, not just words, will be pre-translated into other languages. After all, people say pretty much the same things to each other regardless of the language spoken.

There is the International Bureau of Standards at Sevres, near Paris, that stores such items as the primary kilogram weight. Why should we not, in these days of a shrinking world, have an international standard of the things people say to each other?

This concept takes advantage of the ability of computers to store skill. In order to send an email to anyone in the world, the user will choose the things he wishes to say from lists of words and meanings. It could also be used, of course, simply to translate a message into another language without sending it as an email. On the other side of the world, another user will receive the message and will choose the language for it's display, his own language.

All that is really stored and sent is the numbers, which correspond to the meanings, the input and output displays can be in any desired language. Kathy in New York City, Ngozika in Toronto, Maria in Mexico City, Simon in London, Pierre in Marseilles, Ivan in Gorky, Indira in Mumbai and, Ahmed in Sanaa can all send sets of numbers in the Base Language to each other and receive each correspondence in their own languages. Each will select from lists of pre-translated words and meanings, categorized by parts of speech, to construct their messages.

The email program will have a confirm function before the message is sent. It could contain a function that automatically alphabetizes the lists of words and meanings (which will of course be a different order in different languages). Input will be collected from users in order to improve the program and languages can be added as well as new words and meanings added to existing languages.

A new language can be added as long as someone can speak a listed language as well as the language to be added. The entire concept could be adapted to pre-recorded speech. There will be an override function so that a user can write in something instead of choosing from the listed words and meanings. For example, personal names are not translated. Juan is the Spanish equivalent of John, but if someone named Juan visited an English-speaking land, they would still introduce themselves as Juan and not as John.

I favor listing the words by parts of speech, which are similar in most languages. Speech is broken down into such things as the subject and predicate. Dave is driving, Dave is the subject, 'is driving' is the predicate.

In English it is usually the subject before the verb but not always. A verb is an action. There can be a compound subject or a compound predicate with more than one verb. The predicate revolves around the verb. There must be agreement between subject and verb as far as singular and plural.

There are also so-called "helping verbs", usually the first of a two-worded verb sometimes separated by 'not' or 'never'. One group of helping verbs is the "to be" group including: am, is, are, was, were, be, being, been. Another group of helping verbs is the "to have" group such as: has, have, had. Still another group of helping verbs is the "to do" group, which includes do, does, did. Other helping verbs are: may, might, can, should, could, will, would, must.

A verb has a past, present and, future tense. A past participle is a word such as "seen". The present perfect tense is "have seen". The past perfect tense is "had seen". The future perfect tense is "will have seen". A verb in English that forms it's past tense and past participle be adding -ed, -d or, -t is described as "regular", others are called "irregular" such as "freeze".

The vast majority of words are nouns. A noun is a person, place or, thing. So-called common nouns are no particular person, place or, thing. A proper noun is a particular person, place or, thing and is capitalized.

A so-called comparitive compares two things "bigger" while a superlative compares more than two things "biggest".

Adjectives and adverbs are modifiers. An adjective is a description that modifies a noun or pronoun. An adjective usually answers questions like "what kind of?", "which?" or, "how many?". An adverb modifies a verb, adjective or, another adverb. Adverbs may be formed in English by adding the ending -ly to adjectives. Other adverbs are: a, an, the, this, that, these, those, them.

A pronoun is a substitute for a noun. First person pronouns include: I, mine, my, me, we, our, ours, us. Second person pronouns are: you, your, yours. Third person pronouns are: he, his, him, she, her, hers, it, its, they, their, theirs, them.

Finally, we have the prepositions. A preposition joins a noun or pronoun to another word. Prepositions are: about, above, across, after, against, along, among, around, at, because of, before, behind, below, beneath, beside, between, beyond, by, down, during, except, for, from, in, in front of, inside, into, near, next, of, off, on, out, out of, over, past, through, to, toward, under, underneath, until, up, upon, with, within, without.

In most languages the parts of speech work the same although, of course, they do not combine in the same ways. What we should do is list all the things that people say to each other by parts of speech and assign a number to each that is to be the same in any language. Verbs could be assigned a number beginning with 0, such as 03818. This will include past participles. Helping verbs could be assigned a number beginning with 1. Adjectives, including comparatives and superlatives may be assigned a number beginning with 2. Adverbs would be listed as numbers beginning with 3. Pronouns would start with 4. Prepositions would start with 5. Phrases in a language that would be meaningless if translated word for word could be assigned numbers beginning with 6. The vast majority of all words are nouns and would occupy numbers beginning with 7 through 9. It is possible to categorize nouns in a variety of ways or simply to list them alphabetically.

This would make translation of one language into another simple and easy by making use of the computer's ability to transmit skill.

IDEA #57; THE UNIVERSAL FORMULA: Consider all of the formulas of physics and chemistry. We know the formula for just about any physical process in the universe. Suppose that there is a "big picture" that we are not seeing yet. What I mean is suppose that all the formulas we know are just pieces of a jigsaw puzzle. What if there is a "universal formula" that combines all known formulas of physics and chemistry as well as mathematical operations, since mathematics is a model of the 'real world'?

I do not mean a "theory of everything", such as physicists are now seeking. This would be a combining of all known formulas, seeing which ones cancel out, as pieces of a jigsaw puzzle to come up with a long formula, which may require an entire volume. This would be simply the "Formula of Everything".

IDEA #58; THE GRAV: Many professions have developed their own units of measurement. Sailors and navigators developed the nautical mile for distance on the earth's surface and the fathom for depth. Jewelers use the karat. I wonder why pilots and flyers never came up with units of their own. I thought of a useful unit that would suit anyone who makes a habit of going or sending things up in the air. I believe that all vertical distances from the earth's surface up into the air should be measured in gravs.

One day, while thinking about physics and using a scientific calculator, I came up with what I called "The Grav". The name is short for "gravity". The grav is used in calculations involving falling objects, or the slowing of rising objects due to gravity.

Suppose we want to calculate how long a bomb dropped from an airplane will take to reach the ground? Or how high a cliff is from how long it takes an object dropped from the cliff to reach the ground? We can perform such calculations but I found such calculations tedious. I was sure that there must be an easier way to do calculations involving falling objects.

When something falls from a height, it gains speed at a consistent rate. Considering that we tend to measure time in seconds, there must be a certain distance that would make calculations concerning falling objects easy. On earth, that distance is sixteen feet. A grav is defined as sixteen feet on earth. Since gravity is different on every planet, each has it's own version of a grav.

To find out how long it would take an object to fall from a certain height (discounting air resistance) divide that height by one grav. Then, find the square root of that number. That is how many seconds the fall would take.

You can find out how high an aircraft or structure is if an object falls from it to the ground in a given number of seconds. Simply square the number of seconds. That is the height in gravs.

On earth, a grav is equal to exactly sixteen feet. A grav, of course, is dependent on the units of time in use. A grav amounts to sixteen feet because we tend to use the second as a unit of time. Everyone will find that the use of gravs in calculations involving falling objects will be a lot simpler than otherwise.

Of course as one goes higher in the atmosphere, gravity gradually becomes weaker. However, the grav will equal sixteen feet for at least seven miles altitude or so. This comprises the troposphere where weather takes place and to which almost all aircraft are limited.

Use of the grav as a measurement unit is very appropriate for three reasons. First of all, it is a very natural unit as opposed to being an arbitrary distance like a meter. The grav is based on the rate at which objects fall. Second, outside of atoms, it is gravity that runs the universe, making it logical to base a measurement unit on this all-encompassing force. Third, the new frontier over the past hundred years has been the third dimension, in other words, the sky. Yet, we still think too much in two-dimensional. Use of the grav as a common unit of distance will help to advance our thought patterns toward a full three-dimensional perspective.

When we go to other gravitational environments, such as high altitudes or other planets, the grav will change in length due to the change in gravity. This change however, will be easily measurable and the difference between a "local grav" and an earth grav (16 ft.) will remind us of the ratio of the difference in gravity.

IDEA #59; THE TRIGONOMETRIC PRODUCT: There is a very useful trigonometric function that I have never seen before.

The standard trigonometric functions are ratios of the X-axis, Y-axis and, radius of an angle. The X-axis is the horizontal line, the Y axis is the vertical line and the radius is the line from the origin, which is where the X and Y axes meet, at some angle from 0 degrees to 360 degrees.

The lines of the X and Y-axes thus divide the complete circle into four quadrants of 90 degrees each. Thus, six ratios are formed and these are known as the trigonometric functions, useful for all kinds of measurements.

The sine of an angle is defined as Y/R. When a radius line goes out from the origin point, two straight lines are drawn from any point on the radius. One line goes from the point on the radius to the X-axis, this line must be parallel to the Y-axis. The other line is drawn from the same point on the radius to the Y-axis and must be parallel to the X-axis. The Y of the sine of an angle is the length of the line that intersects the X-axis and is parallel to the Y-axis. The X of the sine of an angle is the length of the line that intersects the Y-axis and is parallel to the X-axis. The R is the length of the radius from the point of origin to the chosen point on the radius.

The sine of any angle is the ratio Y/R. It does not matter what units we measure these lengths with. All we are concerned about is the ratios and no matter which point we choose on the radius or which unit of length we measure in, all results will turn out the same.

Similarly, the cosine of an angle is defined as X/R.

The tangent of an angle is defined as Y/X.

These are the three fundamental trigonometric functions. Then, we have the inverse functions. The cosecant is R/Y, which is the inverse of the sine. The secant is R/X, which is the inverse of the cosine. The cotangent is X/Y, which is the inverse of the tangent. Notice that three of the names of the functions begin with co-, while three do not. The three functions whose names begin with co-get smaller while the angle gets bigger and the other three do the opposite.

Trigonometry has proven to be an extremely useful branch of mathematics when measurements of angles and distances are performed. However in addition to the basic six functions, I believe that there is one more that should be added to calculators and tables because of it's usefulness. My addition I have named the "trigonometric product". The trigonometric product is simply the sine multiplied by the cosine of the given angle. In algebraic terms, Y/R times X/R.

What is different about the trigonometric product in comparison with the other functions is that going from 0 degrees to 90 degrees, for example, the trigonometric function starts from a minimum at 0, reaches a maximum at 45 degrees and goes back to a minimum at 90 degrees. All other functions go from minimum to maximum value over the course of a quadrant. In contrast, the trigonometric product goes from minimum to maximum and back to minimum over the course of the first quadrant.

The trigonometric product is useful for applications in which minimums exist at the beginning and end with the maximum in the middle. Consider a gun as used for artillery or a gun on a warship. If the gun, assuming it's barrel is at ground level with no vertical mounting, will have it's minimum of range if the barrel is pointing either 0 degrees (lying on the ground) or 90 degrees (pointing vertically). The gun will have it's maximum range when it is pointed at 45 degrees (halfway between vertical and horizontal).

The trigonometric product starts out at zero at 0 degrees, reaches a maximum of 0.5 at 45 degrees and, returns to the minimum of zero at 90 degrees. This function can also be used for such things as computing the area of a rectangle. As long as we know what the area of the rectangle would be if it's walls were equal (meaning that the angle of a line joining opposite corners was 45 degrees with either adjoining wall, in other words the rectangle would be a square). The trigonometric product will tell us the area of the rectangle relative to the maximum. All that is necessary is to measure the angle between the line between two opposite corners and an adjoining wall and then find the trigonometric product.

IDEA #60; NEXT GENERATION COMPUTERS: We keep finding ways to improve computers, particularly their speed and storage space. Internet speeds like lightning are available and applications can be found that can make a computer do just about whatever we want it to. The computer is as useful a tool as human beings have ever developed.

But that is just the point. The computer is still a tool. We are improving and will continue to improve computers, even the fundamental way in which computers operate. What we are not doing is stepping ahead to the next generation. No matter what we seem to do with them, computers are still just tools. Wonderful tools they may be, but tools nonetheless.

Think of writing and books as the first generation. Books enable us to store and transmit knowledge. Computers are the second generation. Computers go beyond books and enable us to store and transmit skill. Books can tell you how to do something but cannot do it for you. Computers actually take the skill of others, store it and enable you to benefit from it.

While we are improving computers and making it possible for them to do yet more, we should be increasing the dimensions of the skill that computers can store and transmit. Computers as we know them are one-dimensional, they store and transmit skill but we must tell the computer what to do, on what to use it's skill.

My vision for the next generation of computers is two-dimensional skill. This would make the computer into a companion rather than a tool. Computers may be capable of great things but the initiative still rests with the computer's owner or user. It is the internet that has really made the next generation of two-dimensional or companion computers possible. We just have not caught on yet.

Two-dimensional companion computers need no more technology than is in use now. What is different is that the computer is able to take the initiative. Computer operating systems would contain a program that continuously "learns" about it's user, whether it be an individual, an organization or, a company. The computer is usually never turned off. When it is not in use, it is searching the internet or performing calculations to benefit it's owner. The owner, of course, will be able to override whatever the computer does. The computer will be able to ask it's owner questions in order to better serve. The owner will be able to give the computer input at any time in order to help things along.

The computer will learn it's owner's interests and through continuously searching the internet, performing calculations and, communicating with other computers will offer it's owner new things. If the book was the first generation and computers as we are using them today are the second generation, this would be the third generation. It would also be a classic example of getting more out of the technology we already have as opposed to looking for new technology.

IDEA #61; THE GEOGRAPHICAL FORMULA: This is another formula that every scientific calculator should have. What does one do to find the distance between two points on a map? Usually get a ruler, measure the distance between the points with that and then translate that into the actual distance. But this is a cumbersome process at best and what if we do not have a map containing both places?

I honestly find it amazing that we in the Twenty First Century are measuring distances on maps with rulers. One of the benefits of our system of latitude and longitude is that we can simply and easily measure the distance between two points on the earth using their latitudes and longitudes. Yet, I have never seen this done.

I will not go into an explanation of latitude and longitude here. The Pythagorean Theorem can be modified to work on the surface of a sphere such as the earth. It will not work without modification because although lines of latitude are the same distance apart anywhere on earth, lines of longitude are not. Lines of longitude are the same distance apart as latitude only at the equator. The distance between lines of longitude becomes less and less until at the poles the distance is zero.

The cosine is the trigonometric function that is at 1 for zero degrees a zero for 90 degrees. The Pythagorean Theorem would be modified by multiplying the distance between longitude lines of the two locations by the average cosine of the latitudes of the two places.

2

ALREADY THERE

This is about a tangible object that is already there, in contrast with the first chapter, which was more about mere ideas. Always remember that we can get so much more out of the ideas that we already have. Think of knowledge and technology as an ever-expanding circle like the wave from a stone thrown into a pond. Except in this case, the wave leaves many gaps as it moves outward. Those gaps are the potential benefits that we could be getting but have not noticed yet.

There are certainly many other ways to classify ideas by pattern than I have done in this book. Many ideas are examples of more than one such pattern. A lot of the ideas in this book could have been in more than one chapter. The important thing is that we get used to using patterns to classify ideas. Remember that the patterns and examples in this book are the patterns and examples that my ideas can be categorized as, others may have ideas that can be categorized with a very different set of patterns.

IDEA #62; FOOT PEDAL MOUSE: This works the same as the standard hand-operated computer-mouse but is operated with the foot, like the accelerator and brake in a car. Since just about everyone aged sixteen or above is familiar with the foot pedals in a car, why can this not also be used on computers in the same way as the standard mouse?

The primary beneficiaries of this would be the handicapped, those without full use of the hands. However, it would also add a new dimension to computer games. Such a Foot Pedal Mouse would most likely be constructed the same as the familiar car foot pedals, one side to go forward (the accelerator) and one to stop or go backward (the brake). For children, this would provide familiarity with the principles of driving. In usage, The Foot Pedal Mouse could be combined with the standard hand mouse in a multitude of ways depending on the nature of the game.

IDEA #63; REVERSE PERISCOPE: Submarines use periscopes to see above the water's surface from underwater. Why can we not reverse this to look around underwater from a surface craft or platform?

The Reverse Periscope could be combined with a light and could be directionally controlled. Possibly, it could even provide magnification.

IDEA #64; NON-VEHICULAR USE OF CAR ENGINE: The purpose of a car engine is obviously to provide the power to propel the car. But, we have yet to apply it to the myriad of other tasks to which it could be of use. An internal combustion engine creates a tremendous amount of vacuum. Why can't we come up with some adapters so that in an emergency situation or on occasion, the car engine could be used as a pump while the car was idling?

The top of the car engine, with a few adaptations, could also a fine place to cook breakfast when other accommodations were unavailable. It could even be adapted to cook a meal while driving. The heat, vacuum and, rotary power of the internal combustion engine in every car could be used for far more than just driving. The possibilities are just about unlimited and in most cases would require only a few adaptations and attachments. Once again, this is an example of looking to get more out of the technology we already have as opposed to developing new technology.

IDEA #65; AUTO HEADLIGHTS THAT CAN BE TURNED ON OR OFF ONE AT A TIME: This is one that the producers of James Bond have not thought of yet. Suppose you are following a car at night but you do not want the occupants of the car to get suspicious. The usual way would be to have more than one following car take turns following the target vehicle. To make it easier, why not have a car wired so that the headlights could be turned on one at a time? That way a following car with both headlights, after following the target vehicle for a while, could drop back and turn off a headlight as if it had burned out. Then it could resume following, giving the impression to the occupants of the target vehicle that it was a different vehicle.

Headlights wired in this manner would also offer clandestine signal possibilities. Both headlights on could mean one thing, likewise for one or the other on.

IDEA #66 THE NEXT WAR: Military generals can be brilliant. The problem is that they are often brilliant at fighting the last war. We tend to glamorize heroism in war and this is often counter-productive because it inadvertently glamorizes the way the last war was fought.

Among many examples of fighting the last war are the French in the Second World War. I do not believe that France was weak at all. It is just that when Germany invaded France, the French had glamorized the previous generation of soldiers who had been on the winning side in World War One and now their sons were ready to fight a continuation of that war.

The Germans, in contrast, had nothing to glamorize about the First World War because they had lost. This worked to their advantage. The Germans were fighting a whole new war based on mobility and broad use of air power. The French were planning their defense around massive fortifications such as the Maginot Line, similarly to the way it would have been done in the last war.

Another example is the Americans in Vietnam. It was thought that the Vietnam War would resemble the Korean War with set piece battles and front lines. That was what Americans were used to. Unfortunately, nothing could have been further from the truth. The Vietnamese communists, in particular General Vo Nguyen Giap, redefined the war to suit themselves.

The U.S. military in Vietnam had all the advantages and it looked like a complete mismatch. General Giap knew this and took a different approach. The Americans rarely knew who the enemy was or where he was coming from. Americans were used to a war with front lines but the enemy denied this to them by generally refusing to mass for traditional set piece battles.

As the war dragged on, the typical U.S. soldier wondered what he was doing there, what the war was supposed to accomplish, whether the war was right or who the good guys really were.

As the war dragged on still further, although the Americans were not really being militarily defeated, people at home started to wonder too. In a war without front lines, could we be sure that we were making any progress as year after year went by without an end to the conflict in sight. Did the government know what they were doing? The previous wars had not been like this. In the end, the Vietnamese communists got their way.

Remember that war changes. A war is rarely just like the last war. In fact, no one seems able to define exactly what a war is. Nowadays, they do not even declare war anymore.

I live not far from Fort Niagara. An old fort located on the American side where the Niagara River reaches Lake Ontario. It is a splendid old fort but is of course completely irrelevant today from a military point of view. Today, our military is made up of tanks and airplanes. However, I cannot help thinking that those modern tanks and planes could someday be just rolling and flying Fort Niagaras.

Being big and strong is not all there is to it. America is big and strong but then dinosaurs were big and strong too. The way to defeat a military that is designed to play football is not to play their game but to set the pace and define the conflict so that you play basketball instead of football. All strength has a zone of relevance. The way to handle a strong enemy is to have the conflict outside their zone of relevance. This is what America and it's allies should beware of in the near future.

Remember that the world is changing so fast that no one is sure of the exact definition of a conflict any more. The traditional concept of a victory in battle is not even certain. What would have been considered as a smashing victory in days past may be largely irrelevant today. War is increasingly hazy and fuzzy instead of exact and well defined.

I grew up hearing and reading so much about the Second World War. With all due respect to the veterans of that war, I wish we could put it out of our minds because that is not the kind of war America is ever going to fight again, at least in terms of tactics. (By the way, I am an apocalyptic Christian and know all about the wars at the end of the world.)

My father flew thirty-five bombing missions over Nazi Germany with the RAF. Today, however, I doubt that aerial bombing is as effective as it is made out to be. For one thing, if troops are going into an area that has been bombed, the shell of a building provides ideal hiding places for snipers. Tanks have difficulty moving through rubble. Thick smoke provides cover for movement of the enemy.

America must beware of gaining might and technology but losing simple wisdom. Remember that tanks and planes are only tools and that low-tech efforts can often get the same results as high-tech. Americans like conflicts that are well defined, in which tangible progress is being made and, where there is an end in sight. A clever enemy will deny this to us. Also keep in mind that conflicts may not be against national armies wearing uniforms as in the past. Conflicts may be against shadowy international organizations such as al-Qaeda.

In entering a conflict, do not give away what I call the "situational advantage". This is based on the nature of the conflict itself. If a conflict goes on, which side has time on their side? That side has the situational advantage. America went into the Vietnam War in such a way that all that the enemy had to really do was survive in order to win. In this way, America gave away the situational advantage.

Whichever side has to do the least to win, due to the nature of the conflict, has the situational advantage. In another way, the side that has to be the least vicious has the situational advantage.

The moral high ground is the place to be, the good guys always seem to win, at least eventually. We should strive to be the good guys rather than the masters. This will make things work out much better in the long run. Cruelty is not the way to go and is a sign of inner weakness.

If America is in conflict with Arabs, we usually give them the situational advantage in that they usually know a lot more about us than we know about them, although Saddam handed America the situational advantage in Desert Storm by trying to fight a traditional set piece battle.

Television cameras and reporters will be all over future conflicts and culture is a vital factor in the conflict. People nowadays are very well informed all over the world. If we preach high principles to foreigners, we must live by those principles ourselves. If we are feeding people propaganda or are seeking to buy them or have ulterior motives such as oil acquisition, it will be obvious to them and the outside world. If we tell foreigners that we are trying to help them but show by our actions that we think our lives to be worth more than theirs, they will know. One of the greatest of follies is to truly believe oneself to be above others.

There are many different ways of thinking, ours is only one, I do not apply this concept to religion, I am absolutely a Christian. But it does apply to ways of life in general. For example, Americans tend to think in terms of wealth, Europeans tend to think in terms of quality of life and, Moslems tend to think in terms of morality. Each finds it easy to feel superior to the others while thinking on his terms.

Culture is a part of any conflict in our age of television, the internet and, instant news. If western nations put forth a culture that is rude, lewd, crude and, ignorant then we will attract enemies and repel those who would have been our allies.

This is not just about warfare but, about dealing with the world. We put so much effort into developing new military technology but so little into the old-fashioned idea of learning foreign languages. At the time of this writing, the U.S. soldiers in Iraq have very few people who can speak Arabic. They have the latest military technology but have to find a translator somewhere to ask where to find a bathroom or to question a captive. Maybe language skills should be considered as a vital component of military training.

IDEA #67; DIRECTION FINDER BY STRENGTH OF RADIO SIGNALS: Radio signals and other electromagnetic waves are used extensively for navigation. Why can't we make use of existing radio stations as navigational beacons? AM and FM waves as broadcast by numerous stations everywhere to car radios have different properties, we know that AM waves have much longer range than FM but are blocked by obstacles much more than FM. Transmitters are in fixed, known locations. Obstacles, such as hills and high-tension power lines can assist in determining position.

It is awkward to look at a map while driving and since most car radios already have scanners, all that would be necessary is to measure the strength of a radio signal and match it to a radio map of transmitter locations.

IDEA #68; TIMER USING FLUCTUATIONS FOR SWITCH: This is for use in switches to turn lights on and off for burglar deterrence when residents of a home are away. It could also be used for such things as watering plants. Couple sensors of light, atmospheric pressure and, temperature to turn items such as light switches on and off. This type of system would most likely be calibrated and tested for different geographical locations. Depending on the situation, different parameters can be coupled in different ways.

IDEA #69; PRESSURE CLOTHES DRYER: Clothes can be dried without the necessity of using heat. Air pressure is all that is necessary. Denser air holds more water vapor than less dense air. Put in a load of clothes and compress the air and it will absorb water vapor from the wet clothes. Vent the wet air and pump in dry air, it will absorb more water vapor. Repeat the cycle until the clothes are dry. This would use less energy than conventional dryers using heat.

IDEA #70; OXYGEN GENERATION BY WATER IN LOW-PRESSURE GREENHOUSE: Suppose we wanted to increase the oxygen content in a room. We know that cooler water holds more oxygen than warm water. We know that splashing, such as in a storm, dissolves oxygen in water.

Have a pool of cool and relatively deep water outside. After a storm, circulate the oxygen-rich water into a warm building, such as a greenhouse. Have the water inside moderately heated, possibly by solar heat in a shallow black-painted pond. Oxygen will leave the water and enter the air. If this process is operated continuously, it should significantly increase the oxygen content of the air.

IDEA #71; EXPLOSIVE SHELL TUNES INTO RESONANCE: Of all the high-tech weapons we have today, we are still missing one that would be very effective. An explosion from a shell is not just an explosion. It is also a very loud noise and a shock wave. Military designers have always considered the fire and shrapnel that are the side effects of such explosions, but what about the sound?

We develop bombs and shells but why have we never tried to measure and manipulate the sound of the explosions?

While watching the World War Two movie "The Battle of Britain", I thought of the antiaircraft shells aimed at enemy bomber planes. A bomber plane is a compact standardized structure made of metal. Metal resonates with sound. Any such object resonates at a certain fixed frequency. The best examples are tuning forks and glasses. We have all heard of opera singers who could shatter a glass by holding just the right pitch, which would be the natural resonant frequency of the glass. Likewise in science class a tuning fork that could cause another identical tuning fork some distance away to vibrate.

Suppose the acoustic frequency of the bodies of the Heinkel and Dornier bomber planes raiding London could have been determined. Then suppose that by experimenting with the shape and structure of the antiaircraft shells, as well as the chemical composition of the explosives within, the frequency of the boom when the shell explodes could be manipulated?

Even when a British antiaircraft shell missed it's target, it would cause the structure of any bomber plane to acoustically resonate and vibrate. This could cause extensive damage to the plane. While the British fighter planes, having a different structure with a different acoustic resonance, would not be affected. I believe that this would have been a powerful weapon. Of course, if you are German you would reverse the above example.

This is an example of making use of a byproduct, the sound, which had previously gone to waste.

IDEA #72; AIR PRESSURE ON PLANE TO REPLACE OR BACKUP AILERONS AND RUDDER: There is one thing that I do not understand about jet airplanes. The rudders and ailerons that are used to steer the plane in flight cause drag and thus waste fuel. Jet aircraft that usually fly at high altitudes have pressurized cabins in order to provide normal atmospheric pressure for the crew, passengers and, pilots. Compressed air can hold a lot of force at low weight. Compressed air in tires supports the weight of cars and trucks and compressed air is used in submarines to expel water in the ballast tanks and cause the submarine to rise.

Why not use compressed air from the plane's air compression system to steer the plane?

The standard system of ailerons and rudder operate by slowing one side of the plane down relative to the other side. This produces drag and thus fuel waste. In the compressed air steering system, there would be a hole in the top and bottom of the wings where the ailerons usually are and also in the tail wings. The holes would be the output of compressed air and connected to the air compression system by pneumatic hoses resembling those used to fill a car tire at a gas station.

It would operate much like the release of a balloon. As compressed air is released, it would push the wing up or down and thus steer the plane. Since the plane contains an air pressurization system already, this would not be as big a modification as it would be otherwise

Since the air at the high altitudes where jets usually operate is at low pressure compared to ground level, it would actually not require that much force from the compressed air to steer the plane. In the long run, this would certainly save a fortune on fuel costs. Not to mention that this system is simpler and would be less prone to malfunctions than the current rudder-aileron system.

The only reason that I can think of why this system was not implemented long ago is grooved-in thinking. The rudder and ailerons have been an essential part of a plane's structure going back to not long after the Wright Brothers. When pressurized cabins came into use, this breakthrough was missed.

IDEA #73; DISUSED JET PAINTED FOR PHOTOGRAPH: Sitting in the desert of the southwestern United States are dozens, probably hundreds, of jet airplanes. The planes are parked due to lack of demand. The dry desert climate helps to preserve the planes.

I am sure that the planes' owners wish that there were some way for the planes to be earning money instead of sitting stagnant in the desert.

What about painting the aircraft for hire? Even if a company cannot afford their own jet, wouldn't it be cool to have a framed photo hanging on the wall of a jet with "Joe's Pizzeria special delivery" painted on the jet in the same design as the logo on one of the pizza boxes?

IDEA #74; STORMS ADVERTISE THEMSELVES ON THE RADIO: We use a number of tools to help forecast the weather. Foremost of which is the barometer, which measures atmospheric pressure. High-pressure means good weather, low-pressure means bad weather. When the pressure is falling, a storm is approaching (This is because paradoxically, wet air weighs less than dry air).

Thermometers also give an indication of approaching weather conditions. A drop in temperature raises the relative humidity since cold air can hold less water vapor than warm air and when the air has to release some water, well, forget about the picnic.

One tool that does not seem to have caught on is radio. I do not mean listening to the weather forecast on the radio. I mean that storms actually advertise themselves on the radio. It does not usually show up on FM but have you ever listened to AM radio when a storm is near? I believe that radio should be a powerful tool in weather forecasting. The generation of electricity in cumulonimbus clouds caused by friction between violent currents of air shows up strongly on long-wave radio frequencies such as AM and some short-wave. It would be a simple matter to map such meteorological electrical activity with directional antennas.

IDEA #75; FLASHLIGHT SLIDE PROJECTOR: I believe that one reason that photographic slides are not more popular is the awkwardness and inconvenience of projecting the slides, this could be solved by an adapter constructed to turn a standard flashlight into a slide projector. This would be far easier than the usual slide projectors. The slides could be carried around and shown anywhere almost as easily as photographs.

IDEA #76; REMOTE CONTROL DEVICES: Why, in the era of cell phones, do we need remote control devices for anything? Everything that is done by remote control should be doable by cell phone. A remote control terminal could be an adapted cell phone and a numbering system, similar to voice mail, could determine the function (open garage door, turn off/on lights, enable/disable alarm system) as well as password entry.

And by the way, a cell phone and a calculator have just about the same keys. Why not add a calculator function to a cell phone?

IDEA #77; PNEUMATIC POTENTIAL OF VEHICLE EXHAUST: This makes use of the fact that vehicle exhaust contains the same type of potential energy as pneumatic tools. Of course, gasoline exhaust contains dangerous carbon monoxide and precautions are essential. But for limited use of pneumatic tools in a field situation, the exhaust of a vehicle with engine idling offers a ready source of power. Only a few adaptations like a valve to route the exhaust into the tool rather than the muffler are necessary.

IDEA #78; ELECTRIC LINE KITE: I have long wondered, why not use electricity to make kite flying more interesting? It is necessary to have a string or line going from the ground to the kite anyway. Why not have an electric wire going to the kite? Electric components are light and small and could be mounted on the kite. A source of power, a battery, could be mounted either on the ground or, more conveniently, in the kite-flyer's backpack. Kites could suddenly have all kinds of attachments from lights to cameras to sounds. How about a view of your town from a kite by webcam for a science fair project? What about kite flying after dark with an electric light show?

IDEA #79; BALLOON LIGHT: Suppose temporary overhead lighting is required at an outdoor location. Why not mount lightweight lamps on balloons with the supports for the balloons being a lightweight power cord? I am very surprised that this is not in widespread use.

IDEA #80; PREVENTION OF CAR THEFT BY LOCKING SEAT IN TWO-DOOR CAR IN FORWARD POSITION: There are devices on the market to prevent car theft that work by locking onto the steering wheel. On two-door cars, the seats hinge forward to admit the back seat occupants. Why not construct the seat to be lockable into the forward position? It would be very difficult to steal a car with the seat locked in such a way. It would also make it very difficult to get into the car to steal the radio.

IDEA #81; WINDOWS IN MOVIES: Most people are now familiar with windows as used on computer screens. The next step is windows in movies. This is a classic example of how we keep making improvements (in the quality and special effects in movies) but do not notice the big breakthroughs because of grooved-in thinking.

Think of all movie windows could accomplish. It could show what was going on in other subplots in the meantime instead of the single-window movies that we have now jumping from subplot to subplot. A window could show what a person was thinking without interrupting the main flow of the movie. A twenty-first century movie would have a "main screen" or "main window" and one or more view windows. While one subplot was on the main window, the other subplots could be watched on the view windows. View windows would be created and retired as needed. People with different tastes and interests could emphasize some sub-plots over others.

This use of windows in movies is much more practical for digital movies as opposed to those made on celluloid film. Now that movies are being made digitally, there is no reason to be bound by the limitations of old technology and the grooved-in thinking resulting from obtaining virtually all of our experiences with movie making with this old and limited technology. Also, the fact that television screens are getting larger makes windows more practical. The first movie made with windows will probably be a big hit.

IDEA #82; RADIO INDICATING START OF CAR ENGINE NEARBY: In large cities are many pay parking lots where the driver retrieves his car and goes to a booth to pay for parking privileges according to how long the car has been parked in the lot. Sometimes, things can get busy and car owners have been known to sneak out of the lot with their cars without paying.

The solution is to modify an AM radio to detect and alert the attendant when a car engine is being started. The start of a car engine advertises itself on the radio the same as lightning does. With some experimenting, it could be determined just which frequencies to use and how best to eliminate the effects of nearby lightning. A buzzer could be made to sound any time a car engine was started nearby.

IDEA #83; SPECIAL PARACHUTE SIZES FOR HUMANITARIAN PACKAGES: Parachuted are usually made for one reason, to support paratroopers. The parachutes are thus made in the appropriate size. The only other size of parachute that I have ever heard of is a larger one for vehicles dropped from aircraft.

However, there are other things, such as humanitarian packages, to be dropped from aircraft. We could use a range of parachute sizes, one to fit a load of any weight. There would be no more one-size-fits-all in parachutes. When dropping packages from planes, the parachute could be chosen to fit the package and not vice-versa. This would allow for much more efficiency and flexibility.

IDEA #84; AEROELECTRIC POWER: This is being written just after the great Blackout of 2003. Did you know that a vast amount of electrical power is to be found in the air all around us? I am referring to lightning. I am very surprised that I have not heard of anyone attempting to harness this power. Each lightning strike contains many millions of volts.

Let's use the concept of an automobile battery, which is charged by a current generated by the alternator. Suppose we had tall metal towers with vats of chemicals at the bases of the towers. The vats would basically be large batteries akin to those found in cars. The metal towers would be insulated from the ground but a thick cable would run from the metal into the vat and another cable from the vat into the ground.

Lightning seeking to reach the ground would pass through the vat, which would be designed to store the electrical energy of the lightning as chemical energy. This is how any rechargeable battery works. In many cases, we would not even need to build special towers. Any existing metal in a high place and likely to be hit with lightning would suffice.

Of course, lightning may go from the cloud to the ground or vice versa. It always flows from negative to positive. Sometimes the bottom of a cloud gains a negative charge and the ground below it a positive charge and sometimes it is the other way around. The vat would have to be designed in such as way as to be chargeable from a current passing in either direction. Alternatively, two vats could be present facing in opposite directions. Since electrical charge builds up before a lightning strike, it can be sensed in which direction the current will flow and the appropriate circuit engaged by an automatic switch.

Unlike an auto battery, the vats will require the capability to store the energy of a lightning strike while having received this vast amount of power over a fraction of a second. It would be possible to place heavy-duty coils or capacitors or both in the cable to spread out the energy of the lightning but we must beware that this does not lower the electrical conductivity of the entire system and wastes some of the power or causes the lightning to strike elsewhere, seeking an easier path to ground.

The entire system could be automatic and the captured electrical potential would be stored as chemical energy that could be added to the grid or used for other purposes by simply connecting a switch. A battery provides direct current but this can be converted to alternating current.

I call this concept "aeroelectric power" because lightning actually operates in much the same way as hydroelectric power. A vast amount of static electricity is generated when powerful currents of air rise under large cumulus clouds and brush against opposite currents of air moving downward between the clouds. This movement occurs any time you see fluffy cumulus clouds scattered across the sky. However, it is usually only with the large cumulonimbus storm clouds that it knocks enough electrons out of the atoms in the air currents to build up the charges that produce lightning.

Hydroelectric power generates electricity by using the mechanical energy of falling water to spin turbines. The big difference is, of course, that hydroelectric power is harnessed while aeroelectric power is wasted.

IDEA #85; COMPUTER KEYBOARD SOUND OPERA: Since computers now have decent sound systems, a good way to teach music or make music would be to assign a sound to each key of a keyboard. It would probably be low sound for a lower case letter and a loud sound for a capital letter. This would be excellent for not only teaching but for seeing how a particular song works out.

For beginners, the program would come with sounds pre-assigned to each key. For advanced users the user could assign the sound, depending on the type of music. Being a computer program, the song could be automatically recorded and played back or modified. The user would not need to be able to read music, since it would be known which sounds are assigned to which keys. The volume could be chosen for a note by pressing a number from 0-9 at the same time as the desired key.

IDEA #86; FAST FOOD PAPER BAG MADE OF NAPKIN MATERIAL: The profit margin in most restaurants is pretty narrow. I read that many fast food restaurants are trying to save money on the napkins that are given out to customers. Since the food is given to the customer in a bag if it is a takeout order, why not just make the bag of napkin material? Then the customers could tear off as much napkin material as needed. The cost savings over time would be very significant.

IDEA #87; ROAD MAP ADVERTISING: In these days when everyone seems to be travelling, road maps are more important than ever. I am wondering why businesses have not gotten into the act of making maps or partnering with mapmakers. I cannot think of a better way to advertise than to be located on a road map. Businesses could get together to have a city map printed and have their businesses shown on the map. I am sure that mapmakers could charge businesses a fee to be displayed on a city map. Businesses with many locations, such as restaurants and motels, could have miniaturized logos placed at all their locations on maps and road atlases.

IDEA #88; CORPORATE CURTAINS: So many businesses, especially large corporations, have their own logos. Many larger businesses have their own flags. What I have yet to see is business curtains. The curtains would be a great place to have the company logo or flag. The logo could be made visible from both the inside and outside.

For that matter, it is about time that we could order curtains with a design of our own creation on it. Why can't a photograph or a child's drawing be impressed on the pattern of curtains? How about lampshades too?

IDEA #89; NUMBERED TELEPHONE POLES: This one is long overdue. Every telephone pole in the country should be numbered. This would identify the location as well as the pole. Suppose someone called for help on a cell phone but had no idea where they were? If they could read the number off the nearest telephone pole, that would pinpoint their location.

The telephone pole numbering system could be coordinated with the zip code system. Furthermore, many times addresses are not clearly visible and the local addresses could be incorporated into the telephone pole numbering system. This system could be used to identify any place on any road by which telephone poles run. Which is just about every road.

IDEA #90; MULTI-DISTANCE SIGNS BY SUB-IMAGES: Suppose you are walking toward a sign. Every sign has an effective distance range, depending mostly on the size of the lettering and/or image on the sign. A sign can be made to change as you approach it by the use of sub-images. At a distance, you notice the big picture. But as you get closer, the elements of the big picture could be composed of images of their own, which give a different or follow-up message. These sub-images could themselves be made up of letters giving instructions. The potential of sub-images is virtually unlimited.

IDEA #91; ENCODE STORE RECEIPT MAGNETICALLY: Why in the twenty first century do stores still give out paper receipts for purchases? Don't you think this to be somewhat archaic? We have been using bar codes for decades and it seems that we are doing virtually everything else electronically. Those paper receipts are not only a waste of paper and a source of litter but, how many people actually manage to find a receipt when they have to return something? Why not just use an on/off bit on a magnetic strip on the product to indicate whether the product is purchased or un-purchased? Date and time of purchase as well as location of purchase could also be encoded. It would also be easy to build security features into the process. Paper receipts are really archaic.

IDEA #92; AM RADIO TO TURN ON DEVICE AT NIGHT: Suppose you wish to set an automatic switch to turn on a light or other device during the night but wish to keep it off during the day. The logical thing to do would be to have a light sensor placed outdoors and set the switch to be off when there is the abundant light of daylight and on when it senses the sparse light of nighttime.

But this, for some reason, may not be possible or practical. Another way to do it would be to modify an AM radio. Tune the radio to a distant station that, due to the variations in the ionosphere from day to night, is receivable by night but not by day. When the station is coming in clearly, the switch is on. When it is not, the switch is off.

IDEA #93; LUGGAGE FOR EXERCISE: Many people who travel must leave exercise with weights behind for the duration of the trip. Except for light dumbbells, weights are not usually included in one's luggage. There are exercise gadgets that are compact and portable enough to bring. However, my belief is that luggage itself has potential for exercise during travel. Luggage is usually fairly heavy. All that is really necessary is to adapt the handle of the luggage to accommodate the exercise movements commonly performed by exercisers. It may not be as good as being in the gym but for someone with not enough time to find the nearest health club, luggage exercise is considerably better than nothing.

IDEA #94; STAGE CAR: Have you ever seen people at some event or spectacle standing on their cars to get a view? There is a risk of falling or damaging the car although it does make a suitable temporary platform. The top of a car actually makes a fine temporary stage. Why not modify the roof of a car especially for standing or holding equipment? The top of the trunk could also have a step fitted. The car could either be permanently modified or the platform could be removable. The platform would serve to support weight without damaging the car. If a larger stage is necessary, two cars could be placed side by side. Such a platform would also be ideal for photographers of events.

IDEA #95; ANTIFREEZE STEAM VEHICLE SMOKESCREEN: Once, while driving around the U.S., I was some distance behind an old pickup truck on a rural Tennessee road. Suddenly, the pickup vanished in a cloud of smoke. I had no idea what happened and I sped up to see what I might be able to do to help.

As I entered the cloud of smoke however, there was a scent of antifreeze. It was not smoke at all and the pickup was pulled off the side of the road. It had lost it's antifreeze which, being under high pressure, had instantly vaporized in the atmosphere.

I started to think about the James Bond potential of such an event. If a car wanted help in losing a pursuer, it could sacrifice a certain amount of antifreeze suddenly to create a smokescreen. Of course, it would have an extra supply available.

IDEA #96; CAR'S HEADLIGHT SHOULD INDICATE SPEED: At night, the headlights of a vehicle are not just for the vision of the driver. The headlights also make the vehicle visible. The problem is, it is difficult to judge a vehicle's speed on a dark night by the headlights alone. The apparent angle between the car's light helps to estimate distance but this varies by the type and size of car and is distorted by glare.

Why not mix some colored light into the white light of the headlights? Or, there could be a colored shade that moves across part of the headlight. Why not let blue light represent a slow car while red will indicate a fast car. The light is bluish when the car is at rest or moving slowly and turns redder as the car speeds up. I believe that this should be the next step in vehicle safety development.

IDEA #97; TRAVELLING MOVIES: How to provide entertainment for urban kids on hot summer nights? Try the movie van. This is a van with a projector in back that parks in a suitable location and shows a movie, using the side of a building as a screen.

IDEA #98; RAINWASH POWDER: Falling raindrops actually contain considerable kinetic energy. Why not make some use out of it, starting with washing. Rain already does a considerable amount of washing. It cleans the air as well as dirt and dust from surfaces. Why not help the rain along with a specially made soap or detergent powder. It's first use would probably be to wash cars. But it could be used anywhere that required cleaning and was exposed to the rain. The powder would have to be easily degradable and non-harmful, of course.

IDEA #99; PRE-EXISTANT MAGNETIC FIELDS AS BURGLAR ALARM: Inside virtually any building is magnetic fields from such things as electrical appliances. This makes possible a less obvious, less intrusive and, less expensive burglar alarm system. The system would only have to be calibrated to the normal pattern and set to trigger an alarm at any disturbance. A human body moving through a magnetic field would register as a disturbance.

IDEA #100; NUMBERED TRAFFIC LIGHTS: In the automobile age, traffic lights are the focal points of cities. Why are not traffic lights shown on maps? Traffic lights always seem to figure prominently in directions. "Go east three lights, make a left, go two lights and there it is". So why not assign numbers to traffic lights? This simple idea would be a great breakthrough in urban navigation. Once again, we can get so much more out of things as they are before we even develop new technologies.

When you are in an unfamiliar area and you stop at a traffic light, why do you have to look around to find the street signs to see which intersection it is? Why not put street signs on top of the traffic lights in addition to the sides of the road?

IDEA #101; ELECTRIC PYLON ADVERTISING SPACE: What about those metal towers carrying electric lines that seem to be everywhere? Why put up billboards along the roads when the towers are nearby? I often see billboards near electric pylons. Why not just combine the two? The electric company could sell space on the pylons for advertising. This would lower the cost of advertising, raise money for the electric company so that they could lower their rates, and conserve the environment by lowering the number of billboards. This should have been done long ago.

IDEA #102; MICROWAVABLE HOT WATER BOTTLE: On those freezing cold northern winter nights a hot water bottle creates a nice, warm bed. The trouble is that is a hassle to boil the water and pour it in the bottle. What we need is a microwave-able hot water bottle? It would be so much easier and would certainly increase hot water bottle use. So many other things have been adapted for the microwave era, why not the old hot water bottle.

IDEA #103; ANIMATION FROM PHOTOGRAPH: So much of the entertainment world is cartoons. Why not make cartoons from photographs? Instead of just hanging a photo on the wall of someone, why not have a cartoon of the person too? Giving an animation, as opposed to a photograph, would make a nice gift. All that would be necessary is to scan a photograph, reduce the number of colors and, smooth it over. Given the prevalence of lossy technology, creating cartoon images from photographic images would be easy.

IDEA #104; IMAGE ON VENETIAN BLINDS AND ENVELOPES: Why are venetian blinds so blank? I cannot even remember seeing a decorative pattern on venetian blinds, just the same old white. It should be easy to imprint a photograph on venetian blinds. Just imagine your home being the first on the block with a family photo imprinted on your venetian blinds instead of the usual blank white.

For that matter why can you not buy custom-made envelopes with the image of your choice on it from a photograph or other illustration? Then, you can send a letter to someone not with a photo enclosed but with a photographic image of your choice on the envelope.

IDEA #105; IDENTIFICATION BY SOUND ABSORPTION: It is possible to identify and classify people by the body's absorption of sound. Any body will absorb and reflect certain frequencies more than others. A sonic profile can be made of a person in a room. It is even possible to use background noise or music to classify a person by sound absorption. If a series of sounds of differing frequencies can be emitted and received after passing by a body, a unique sonic profile can be established. This would, in most cases be useful as an alternative to or supplement to visual cameras.

3

MARRIAGES

Marriages are combining more than one item. It is the art of killing two birds with one stone. It differs from the previous chapter in that the two items in marriages are more or less "equal". In 'Already There' we usually has a primary idea that spawned secondary ideas. All multi-purpose gadgets in stores are, of course, marriages.

Part of the problem is that we are so specialized. Ideas come from different companies and society is run by many government agencies. This prevents us from seeing many possible marriages. Imagine that there was one company that made everything and it was government run and so had control over things like the phone and mail systems. No, I am not suggesting some new super socialism but, we would see so many marriages that we do not see now.

IDEA #106;COORDINATION OF ZIP CODES AND AREA CODES AND ABILITY TO TELL TIME ZONE BY AREA CODE: At present in the United States, there is no coordination between the zip codes used in the mail and the area codes used in the telephone system. I believe that this is a colossal waste of efficiency.

We live in a world that tries to make things easier to navigate and understand the things of daily life. The area code is three numbers and the area code is five numbers. Why not coordinate the two so that if you know the zip code of an address, you will also know it's area code without having to find a way to look it up?

There should be no such place where the zip code is 14304 but the area code is 716. Make the area code three of the five numbers of the zip code. This would create a system in which a zip code would be 1% of an area code. The area code would be either the first three numbers of the zip code or the last three. Thus, it would be 71604 or 14716 instead of 716 and 14304. Except for momentum, there is really no reason for lack of coordination between these two systems of communication.

For another thing, it is increasingly easy to call anywhere in the country or for that matter, the world. I believe that a caller should be able to tell the time where he/she is calling by the area code inside the U.S. or Canada and the country code outside North America. This really should have been done from the beginning. But phone calls to faraway locations were not common then. There should be no reason to have to look up the time difference before making a long-distance call.

IDEA #107; THE BLURSCOPE: This is a cross between a telescope and a camera with many different applications. It works by combining an exposure meter with shutter speed. If we can tell at what speed the image is clear, we can tell the speed at which an object such as a wheel or turbine is spinning.

IDEA #108; BICYCLE PUMP SUBMARINE: A submarine operates by taking on water into a space between two hulls to make it sink and then using compressed air to force the water out and refill the space between the hulls with air when the sub is to surface.

I believe that such a setup in miniature would make an excellent and educational toy for an older child. A chamber in the toy sub would be filled with compressed air prior to launch with a simple bicycle pump. Such a pump could even be built into the sub. Radio control probably would not be as easy as with conventional radio-controlled airplanes and cars because higher frequency radio waves do not travel well through water. So, a course for the sub could be pre-programmed, including the submersion and resurfacing points.

The sub would probably have a conning tower and fins just like a regular sub and could have a strengthened front for accidental impacts and a cage around it's propeller. Like non-nuclear submarines, it would be driven by a battery-powered electric motor. The sub's pre-programming function could have a pressure meter to measure depth underwater as well as a rudder for steering. The guidance system could have a number of possible journeys already pre-programmed and the user could select one.

It is time for submarines to join the remote controlled airplanes, helicopters and, boats now used by hobbyists.

IDEA 109; CASSETTE RECORDER TECHNOLOGY FOR AUTOMATIC PILOT: There are potentially many more uses for inexpensive cassette players, primarily as pre-programmed guidance systems. "Beeps" could be recorded on a cassette tape at a high enough threshold so that static could be ignored.

The tape would be played in the same way as if music was playing but when it came to a beep, some other function would be performed such as turning a camera, rotating a cam or, changing a switch of some description. The electric emanation from a beep would act activate a relay switch instead of being played audibly by a speaker. Possibly there could be multiple "tones" recorded and a number of switches or other control devices each responding to it's own tone, in the same manner as is used in telephones.

This is not a high-tech idea but it's great advantage would be affordability. This idea need not be confined to cassette tapes. A computer-generated sequence of beeps on a DAT tape could store a very complex plan.

IDEA #110; ELECTRIC CAMERA FOR KITES: Since we have today such miniaturized cameras for computers and webcams, why not use them to make kite-flying more interesting with a kite-mounted camera? If we had a fine electric wire in place of the conventional kite string, it would be simple to operate the camera from the ground electrically. This would combine kite flying and photography. The photo could be taken digitally, without using film. The camera could be weighted to point as straight down as possible and could act as the kite's tail.

IDEA #111; AIR POLLUTION TESTING BY USE OF LOCAL LIGHTS AND STARS: Astronomical observatories are located to avoid air pollution as much as possible. But why couldn't we use the same type of equipment and techniques to test for air pollution? Most measurements of air pollution are taken by sampling and this would greatly increase the range.

Stars give off a characteristic spectrum and standard lights also emit characteristic frequencies and intensities of light. Testing can be done to determine which pollutants have what effect on existing spectra of stars and artificial lights. This makes it possible to take air pollution samples by examining the spectra of lights or stars and then making a comparison with the known spectrum (when pollution is absent).

IDEA #112; TALKING THROUGH LIGHTS: This would be at least an interesting science fair project. Fiber optic cables use light to carry hundreds of conversations at once. Why not use light from a conventional lamp of some type to transmit a voice? Visible light is of extremely high frequency relative to radio waves and could easily transmit a conversation with a simple apparatus. An ordinary resistance microphone in series with the lamp would cause the light to transmit speech by amplitude modulation. At the receiving end, a light meter focused on the transmitting light would operate a relay switch to a speaker.

This would be an ideal means of communication when the conversation is to be kept secret. Light can be made much more directional than radio waves and the message much more difficult to intercept, even if the bad guys knew that we were going to communicate by light beam. Since there are lights everywhere, a portable microphone could be easily plugged in to any available light that was in view of the receiving party.

IDEA #113; WATCH MEASURING PULSE USING WATCHBAND: Taking the pulse is considered as the quickest way to measure heartbeat during exertion. How many times have you seen joggers taking their pulses the old fashioned way while wearing watches? Why not construct a watch that can take pulse measurements? A watchband and a meter device (the watch) is already present. All that is necessary is to add a pressure sensor and a display. This would make possible the easy taking of the pulse anytime and would make the wearer of the watch much more aware of their pulse and cardiovascular fitness level.

IDEA #114; CARBON MONOXIDE SENSORS ON STREETLIGHTS: All over the world webcams are being placed on busy streets and intersections so anyone can see the traffic flow and volume in real time. However, I think that it would be easier to mount carbon monoxide sensors on telephone poles and determine traffic volume based on measurement of carbon monoxide. CO from an auto exhaust is hot and so rises rather than remaining at ground level. CO sensors are inexpensive and telephone wires are present for communication of the data.

The information could be accessible to anyone and could be used for large-scale data gathering and environmental use. The data may be useful in ways that we do not even realize yet.

IDEA #115; MEASUREMENT OF ALTITUDE IN HILLY COUNTRY TO DETERMINE LOCATION: I believe that the presence of hills should be an asset to navigation rather than a hindrance. It is true that hills obstruct the view of the sky and with certain minerals present can interfere with magnetic compasses. On the other hand, any point on a hill shares it's altitude with only a limited number of other points in the area. Since altitude can easily be measured by pressure, this gives us a powerful navigational tool.

If we know the pressure at sea level, or at a convenient ground level, we can determine our altitude anywhere we can measure air pressure. Many cars have a built-in compass. Why not couple it with an altimeter? If maps were marked with altitude contour lines, whenever you were lost it would only be necessary to take an altitude reading and find the contour lines corresponding to that altitude. This alone would probably not tell you exactly where you were but you would know that you must be somewhere on that contour line on the map.

IDEA #116; CAR LIGHTS AS SURVEYING BASELINE: People go places in vehicles to accomplish all kinds of tasks but the designers of vehicles seem to think only of transportation issues and comfort. I believe that cars have the potential to assist in a wide variety of chores with simple modifications. We could even have a "utility car" which could be the vehicular version of a Swiss Army Knife. The car itself could be much more of an all-purpose machine. One of those chores is surveying.

The headlights of a car contain powerful beams and are spaced a known distance apart. Suppose we point the car at a distant surface such as a wall. If vertical parallel slit masks could be placed over each headlight and one of the lights could be slowly angled until it's vertical light slit exactly covered the other light slit, it could easily be determined the distance to the distant surface. The calculation could be done by triangulation since we know the distance between the vertical light slits and all the angles involved.

IDEA #117; PHOTOGRAPHS AND LAMPSHADES: An ideal place for a photograph is to make it into a lampshade. There are many things that can be done with a photograph but I have never seen one made into a lampshade.

IDEA #118; COMBINE TOOTHPASTE AND TOOTHBRUSH: We have disposable cameras, why not toothbrushes? A toothbrush should be changed at about the same time as a tube of toothpaste anyway. It would be a convenience and space-saver. Both items could be in a tube that could be disjointed, toothpaste applied and then re-jointed. Or, toothpaste could be inserted into the toothbrush from below through small holes via a twisting mechanism.

IDEA #119; JACKET SNAPS INTO CARRY BAG: A jacket or a carry bag can be made from about the same amount of material. A jacket being worn has a close similarity to a full bag being carried. Combining the two would be the ultimate in utility. Snaps or buttons on the jacket can also button the bag. Sleeves on the jacket can join to form a carrying strap on the bag.

IDEA #120; INFORMATION BOOTHS AMONG TOLLBOOTHS: A toll plaza is often the beginning or end of a highway journey. A toll plaza is actually an ideal place to ask for information. People ask for directions and advice at tollbooths anyway. I do not see why the booths could not sell maps and newspapers also.

IDEA #121; COMBINING HAND TOOLS WITH LIGHT: It is often necessary to use tools in the dark or in dim light, especially in certain work environments. In such cases, it is usually necessary to have someone hold a flashlight while the other person uses the tool. Why not include a source of light in the tool, most likely in the handle. It could contain a rugged light that could be turned on when necessary or could be made of material that would absorb light and then reradiate it.

IDEA #122; FUEL POWERED PISTON GUN: Since we have not yet found a way to do without wars, we can at least find a way to avoid storage of dangerous ammunition. Shells or bullets could be propelled by a controlled gasoline burst, in exactly the same way that the pistons in an engine are driven. The gun would be an engine piston with a barrel. It would operate in much the same way as an engine piston, with input and output valves for gasoline vapor and exhaust and a spark plug operated by a trigger.

Probably the only reason that this has not been implemented yet is that guns have been around a lot longer than internal combustion engines. By the time that engines and gasoline came on the scene, the tradition of gun-making using gunpowder had been around for centuries.

IDEA #123; ALARM CLOCK AND LAMP COMBINATION: Alarm clocks and bedroom lamps go together like a hand in a glove. Radios combined with alarm clocks have been around for decades. The combination of the alarm clock with a lamp is an even more sensible combination. The base of the lamp would contain the alarm clock. The alarm clock could be battery-powered to avoid time loss in case of power outages.

IDEA #124; GEOGRAPHY AS SHAPE DEFINITIONS: Geometry only defines basic shapes such as circles, squares and, rectangles. Most of the rest of all possible shapes is defined simply as "irregular".

What about geography, in a world atlas just about any two-dimensional shape can be found in the shapes of nations and states. Irregular geographical shapes can be defined by the similarity to a geographical entity. There are approximately 220 nations in the world and 50 U.S. states with a wide variety of shapes. This process would also be useful for learning geography and I used to use this system to keep track of irregularly shaped scraps of fiber when working in a factory.

IDEA# 125; COMBINE MODEM AND SOUND CARD: Computers have both a modem and a sound card. Both use a UART chip to convert signals between analog and digital. The two are not in use at the same time. Why not combine the two?

IDEA #126; COMBINATION FIREPLACE AND BREAD OVEN: The taste of home-baked bread is legendary, although it is rarely done nowadays. A bread oven is not too different in structure from a fireplace. I think that this would make a delicious combination.

IDEA #127; POTATO AND PIZZA: I just cannot get rid of the idea that there is a place for potato on pizza. I do not pretend to be a chef but I know a good thing when I see it. Potato in some form would make a piece of pizza really filling. Potato is inexpensive and it would be easy to come up with ways in which to apply it. Rice would also make another possible complement to pizza.

IDEA #128; THE TRUCK BUS: Vehicles tend to be designed as either a truck or a bus but not both. But there is much potential overlap that is rarely considered. In many nations are trucks sitting idle while buses are needed or vice versa.

What is needed is a truck container for seating passengers. The container could be hitched to the truck like any freight container. It would also be possible to design a bus with removable seating for use in carrying freight, although I believe that the first solution is probably more practical.

IDEA #129; COMBINATION PEN AND COUNTER: It is often necessary to make a count of something without miscounting or losing count. Why cannot an ordinary pen be used as a counter? There are colored pens with several different color modes. A pen should be able to be used as a counter simply by clicking. It is somewhat wasteful that something like a portable and widely used pen is used for only a single task.

IDEA #130; LIGHT DOUBLING AS SLIDE PROJECTOR: When one takes photographs with film, there is the choice to have the film made into either prints or slides upon development. The vast majority of the time, the choice is prints. I like slide shows but the downside is the awkwardness and inconvenience.

Let's make it easier. Instead of those old slide projectors, we need a simple device to fit over a light and so turn any lamp into a slide projector.

IDEA #131; INTERIOR CAR LIGHT AS MAP SLIDE PROJECTOR: It is difficult for a driver to read a road map and drive the car at the same time. The interior light could be fitted as a small slide projector to project an image onto a flat area on top of the dashboard. Maps could be made as slides. Initially, the standard fold out maps could contain a pocket holding the same map as a slide. Either the entire map or a portion thereof could be projected by a variable distance focus on the light. Navigation would be much easier.

IDEA #132; TELEPHONE RING FOR OTHER REASONS: This is a good example of gadget unity. Why does the telephone ring only when someone is trying to call you? The telephone and it's ringer would make a wonderful alarm. There could be different rings for phone and non-phone functions. There could even be a pre-recorded statement as to the reason for the ring. This would be useful for just about anything that an alarm or audible notification would be used for. Anything from a doorbell to the clothes in the drier being done.

IDEA #133; INTERNAL COMBUSTION ENGINE LIGHT: Suppose a powerful source of light is needed. Consider the standard internal combustion engine not just for the motion and heat it produces but also light. Inside the cylinders of every engine is an intense but never seen light.

Today we have super strong glass that can take blast furnace heat. If the cylinders and part of the engine block could be made of such glass, we would have an extremely powerful source of light. It could be constructed as either a spotlight or a floodlight. On an airplane engine, for example, the engine light would produce a powerful beam to search for shipwreck survivors below.

IDEA # 134; REVERSING ESCALATORS FOR EXERCISE: In northern climates, a major challenge is finding a place to get a cardiovascular workout in winter. People sometimes go to malls to walk briskly. An escalator reversal for running would be ideal. If there was more than one escalator, there could be more than one speed available.

IDEA # 135; SURGE PROTECTOR BUILT INTO COMPUTER: Why does everyone have to get a surge protector when they buy a computer? Why not just build the surge protector into the computer?

IDEA # 136; UNIFICATION THEORY FOR ACADEMICS AND ATHLET-ICS: Sports play an integral role in development of youngsters. Athletics develop fitness, motor skills, reflexes, concentration and, coordination as well as awareness of the body's abilities and limitations. Sports are the primary venue for the development of teamwork. The fitting in of new people on the team and, becoming accustomed to performing in front of others and being evaluated by peers is accomplished on the playing field.

Sports provide background skills that are very difficult to impart in the classroom. Throwing a ball, for example, involves complex calculations. With a baseball, the arm is the power and the wrist and fingers provide the fine control. The entire procedure requires estimation of distance as well as vertical and horizontal angles.

Instruments, such as a bat and glove, are usually a youngster's first experience in using a tool as an extension of self. Running requires choice of the best stride for efficiency. Keeping score is an introduction to applied mathematics, particularly if records and averages are kept.

Such intangible education is not, of course, limited to competitive sports. A bicycle is one of the finest possible introductions to the mathematical relationship between angular and linear distances, the angular distance in the wheels and the corresponding linear distance the bike has traveled.

Sports develop decision-making skills and problem solving. A young athlete learns to evaluate odds and risks in ways that are difficult to provide in the classroom. Sports teach the concepts of strategy and strategic positioning by trial and error. Throwing a ball at a moving target nurtures a sense of intrapolation and extrapolation.

The patterns found in sports are the same ones that occur all throughout life. Offense and defense, sometimes we are on an offensive roll and, sometimes we are on the defensive. Being up at bat teaches one to learn to make the most of an opportunity. In all sports, the objective is to either make things happen or to stop things from happening. This is just the way it is in life as a whole. Sports, like life in general, offers windows of opportunity in space and time and both consist of aiming at targets. The patterns are all the same; what you are aiming, what you are aiming at, what is in between.

The intangible education to be gained through sports is a necessary complement to the tangible classroom education simply because it offers so much that cannot be found in the classroom. The complex intuitive calculations, the movements at relative speeds toward windows of opportunity are nearly impossible to replicate in the classroom in such an effective way.

What better preparation is there for being a good automobile driver than the judgment of distances and angles and windows of opportunity on the playing field? How can students learn in a classroom how to hit a moving target, get around an obstacle and, evaluate multiple causes and effects as well as on the playing field? The rules of simple childhood sports are the best introduction to the concepts of law. Disputed calls on the baseball diamond mirror the cases in the courts in the adult world. For a game to have meaning, a set of rules must be agreed on and must be obeyed by both sides. Some kind of "law enforcement" as well as "justice" is required, even if there is no assigned referee.

A child's introduction to leadership and organization usually comes as organizing and selecting teams for a sports event. Setting up a playing area is a mirror of architecture and urban planning.

The items involved in sports may well be a future scientist's introduction to the concept of materials. Balls are made of rubber, leather and, plastic. Bats are made of aluminum and wood. Inflatable balls involve air pressure. Things made of different materials behave in different ways. The rough surface of a basketball involves friction.

The entire concept of sports on a playing field imparts an appreciation of the outdoors. The physical exertion involved in sports increases awareness of the body and interest in fitness. The inevitable minor injuries provide a sense of both toughness and safety.

Simple athletic activities such as hitting or throwing a ball and running make use of the body as a high precision machine. Throwing a ball at a target is usually a youngster's first experience with steering. Use of a baseball bat involves detailed calculations with leverage and angles, the direction in which the ball travels depends upon where on the bat it strikes.

Of course, not all such intangible and informal education involves competitive sports. Toys allow a child to become familiar with a replica of the world in miniature. A yo-yo will accustom a child to the belts, pulleys, chains and, gears that are found in such a wide variety of machines. A child who builds a tree house may grow up to build skyscrapers. A tricycle leads to a bicycle that leads to a car. Handling playing cards accustoms a child to handle stocks and bonds. Seesaws provide a good introduction to any machine involving a counterbalance. Sliding down a slide does the same for drainage systems and canals

Sports have long been considered as a good preparation for military service. The Duke of Wellington said that the Battle of Waterloo was won on the playing fields of Eton. Playing with toy guns are training for the future infantryman and playing chess or with toy soldiers is an exercise for the future general.

Athletic competition against a similar group in a different uniform is, of course, what warfare is about. Scouts and fraternities are further pre-military preparation. From summer camp to boot camp. Covering bases is training for guard duty and the line in a football game follows the same pattern as the front line in a conflict. Learning how to handle a bat comes before handling a rifle and throwing a ball precedes throwing a grenade. Dealing with the physics of flying footballs, basketballs and, volleyballs provides a feeling for the ballistics of bombs and artillery shells.

Sports as a whole are a model of life in general. Any team sport involves a variety of diverse tasks and the young athlete can find the one he is best suited to. The entire team represents different positions and talents working together. Changing from one sport to another mirrors career changes in adult life. The colors, banners and, logos are a preliminary to the corporate logos and national flags of adulthood.

The problem is that sports are considered as being for recreation only by much of the education community. The physical prowess provided by athletics is acknowledged but the informal and intangible academic foundation is not. I am certain that the academic performance of students can be significantly improved by making the most of sports. Athletics have always played a vital role in education as we have seen. I want it to play even more of a role.

How could America idolize Joe DiMaggio but get caught off guard by Sputnik? The patterns and physics in a fly ball and a satellite in orbit are virtually identical. The trouble is that adults are too specialized and organized. I noticed in high school that the gym teachers usually hung out together and there did not seem to be a lot of interaction with the academic teachers.

A student could be either a "jock" or a "brain" but rarely was both. Why does a jock have to be a "dumb jock", while someone who is smart must be a nerd and would rarely be an athlete. What I am formulating here is a new direction for the education community. Removing the artificial wall between academics and athletics is the way to improve academic performance. This is especially relevant now that many more girls are playing sports.

I believe that sports can be far more beneficial in education than the examples that I have given above. A jock should, in fact, have a big head start in the classroom. Standard competitive sports involve all of the same physics and mathematical patterns as the school textbook does.

Sports are not something "different" from academics but a "lab" of the things that will be taught from the textbook. We need to start realizing this. I believe that this was realized more in the past but with our over-specialization today, it has largely been forgotten.

Sports build an excellent foundation for mechanics. The human body, which is used in athletics, is the ultimate machine. One's first encounter with the concept of mechanical efficiency may be the choice of the best stride for running. The throwing of a ball at a moving target is often a future engineer's first use of precision measurement. The two basic methods of propulsion are both widely used in sports; the ballistics of throwing and the action and reaction of running.

A spinning ball demonstrates the basic ballistic principle that a spinning object is much more likely to move in a straight line than one that is not spinning. This is because the spin averages out the object's center of gravity. Throwing a ball in a wind serves as an example of the aeronautical concepts of air speed and ground speed. The stitching on a baseball or football to facilitate throwing is an example of mechanical cogs.

Basketball is a sport of great precision. The precision of the shot is far more important than it's power. Precision bouncing is also a vital part of basketball. How is it possible for a basketball star to not also be a star in precision technology? Maybe this would be the way it is if we would break down this destructive artificial cultural wall that we have built between sports and academics. There are, of course, sports examples used in textbooks but we can get far more benefit out of the similarity of the patterns in athletics and academics than this.

A hit or thrown ball has many textbooks of lessons in the behavior of matter and energy. A volleyball sailing over a net makes use of exactly the same physics as a satellite in orbit. When it comes to mathematics, a ball represents the dependent variable while gravity represents the independent variable. The stars on the playing field should also be the stars of the classroom.

In baseball, the hitting of the ball converts the radial motion of the batter into the linear motion of the baseball. This same conversion between radial and linear is also used in hydroelectric generation and in internal combustion engines. A baseball striking the bat involves the same mathematical concepts of angles and axes that any gear system does.

The legs work together when running in the same way as do the pistons in an internal combustion engine. The dribbling of a basketball while moving follows the same pattern as the timing of an internal combustion engine. Jumping rope or using a swing follows the same pattern as the coordination of the crankshaft and camshaft in an engine. The dribbling of a basketball or the carrying of a football to best keep the ball away from opponents mirrors the designing of a car to best protect the passengers in a crash.

Baseball also relates well to electricity. The force of the hit or throw can be compared to the voltage. The ball itself represents the amperage. The air resistance and gravity is congruent to the resistance. The running of the bases is a model of an electrical circuit. Home plate compares to the battery or generator in the same way that the bases compare to light bulbs. The power flows out through the load and returns to the other terminal on the source.

Sports are a veritable lab of the physics of energy. The hit of a baseball and bat is one of the best examples of the addition of energy of collision. The bounce of a ball shows how energy is not created or destroyed but merely transformed from one form or direction to another. A bouncing ball illustrates entropy, the decrease of energy in an object. Each bounce has less energy than the previous bounce. The ultimate halt of a rolling ball is another good example of entropy as well as friction.

The hitting of a ball with a bat involves a complex interplay of forces, momentum, gravity and, angles. A ball bouncing around the gym is a better lab of matter, energy and, angles than can possibly be found in the classroom. Each bounce displays elasticity, vectors and, reflection.

Any hit or thrown ball has a textbook of lessons in the behavior of matter and energy. It contributes to an understanding of gravity, it is a conservative force meaning that the work done does not depend on the path of the object but only on the start and end positions. It also throws a light on momentum, an object in motion keeps moving until stopped by an outside force, and the classical mechanics defined by Newton's laws.

All of electricity and magnetism can be represented by Maxwell's few equations, which are analogous to Newton's laws. A simple game of catch as opposed to a game of baseball illustrates the difference between scalar and vector. A spinning ball or Frisbee displays the concept of spin around an axis that is so important in the universe. It shows a youngster how an object can possess angular and linear momentum at the same time.

An athlete has quite a head start on understanding science and mathematics. So how have we managed to separate athletics and academics to the point where most students are considered as proficient at one or the other but not both? Something is not right here. We are moving in the wrong direction, let's move back in the other direction. There should be a close unity between athletics and academics.

The forces determining the behavior of a ball are the same ones governing atoms and planets. And just as we have different size balls for different sports, we have different atoms for different elements.

In fact, sports are mirrored on the atomic scale just as much as on the planetary scale. If an individual athlete can be compared to an atom, then a sports team can be compared to a molecule. There are many possible combinations of players on the team, which varies the "chemistry" of the team. Sports teams, like metal alloys or any chemical compound, are made of individuals with varying properties. A sports event can actually be compared very well to a chemical reaction. The pairing of two teams is analogous to combustion.

There are three states of matter; solids, liquids and, gases, with the state determined by the strength of the bonds between the atoms. This is mirrored in sports. There are sports with fixed positions such as baseball, which corresponds to solids. There are sports with more flexibility in the positions such as football, which is analogous to liquids. Finally, there are sports with still more flexibility in positions such as basketball, which corresponds to gases. A young athlete should thus gain a head start in chemistry.

The ball is the center of a game just as the nucleus is the center of an atom. Large and small balls are analogous to hadrons and leptons. Baseball fields have long been compared to atoms. The infield resembles the lower electron orbitals in the same way that the outfield resembles the outer orbitals. Electromagnetic waves, such as light, are produced when an electron drops from a higher to a lower orbital and thus releases energy.

Waves and cycles figure very prominently in the universe and fortunately, this is mirrored in sports. A rising and falling ball shows the young athlete and budding scientist the concepts of wavelength and amplitude. Swings and rope jumping illustrate cycles and harmonic motion. The dribbling of a basketball is probably the finest introduction to rhythm, not only to be found in music but all over nature.

The behavior of light and other electromagnetic waves is also reflected in sports. Bouncing a ball off a wall or a basketball off a backboard follows the same concept as light reflecting in a mirror. Just as the curved surface of a baseball bat sends a baseball off at an angle, a curved lens or mirror refracts light. A clean hit in baseball can be compared to specular reflection in the same way that a foul ball can be compared to the scattering of light from a rough surface.

Sports can provide an excellent foundation for understanding the workings of nature. A model of the relentless competition and the predator and prey relationships of nature is found in competitive sports. The seeking and evading, the deceptions and captures, the development of tricks and strategies are to be found both in the wild and on the playing field.

Mathematics exists because it provides a model for nature and the universe. If sports accomplish the same thing, then we should expect some congruence between sports and mathematics and every accomplished athlete should be just as accomplished in mathematics.

Any sports action involves very complex calculations, even if those calculations are done intuitively. The keeping of scores, particularly baseball stats and bowling scores, further enhance the relationship between sports and mathematics. The operation of an elimination tournament is a fine introduction to logic and is how many computer languages work, such as the "if, then…" of BASIC.

A coin toss is the first step in understanding statistics. How a sports team is arranged is known in mathematical terms as a permutation, each different possible arrangement of a number of things. All sports events involve intangible mathematics such as evaluating odds and risks. The reason a spinning ball will move in a straighter line is related to the concept of averages, any differences in mass over the surface of the ball "average out". Plainly and simply, the determination of where a hit or thrown ball will land involves the same concepts of vectors, angles and, forces that run the whole universe.

Facing an opposing sports team is similar to facing an algebraic equation. There are some known facts and some unknowns, often represented by "x" in algebra. There are dependent and independent variables that will determine what the opposing team will do. The team's intention is an "x" for now. The team's actions will be determined by a "function", a dependent variable depending on some independent variable. Just as in algebra, we would say that 3x-7 is a function of x.

Athletics is a feast of geometry. Every bounce of a ball is, of course, a complex geometric problem. Any hit or thrown ball is on a course expressible by a horizontal and a vertical angle. Throwing a ball at a moving target forms a triangle and should give a youngster an idea what the trigonometric functions are about. A few games of table tennis should have a student ready for trigonometry class. In the running of bases, each base represents a quadrant.

In fact, I would go as far as to say that the catching or interception of a ball is basically a trigonometric calculation. Wayne Gretzky was known as the "master of angles" and every hockey player should be a master of geometry and trigonometry. The numbered yardage on a football field should make a student familiar with analytical geometry. Rope jumping should, in a similar way, make a student comfortable with shapes in analytical geometry such as sine waves. The dribbling of a basketball also conforms to the same pattern as a sine wave.

Calculus is used for such tasks as predicting trajectories, economic cycles and, population growth. A ball is thrown and then it falls. Gravity is the constant, the force and angle at which the ball is thrown is the variable. The curve of a bat that hits a ball is a good example of the "instantaneous curve" concept in calculus. The balancing required to ride a bicycle should also give a student a feel for calculus.

When we want to throw a ball to someone a certain distance away, we must throw the ball at a certain angle into the sky in order to do so. If we consider the horizontal as the x dimension and the vertical as the y dimension, when we throw or hit a ball achieving x will bring negative y (-y) so, we introduce positive y and when the baseball falls and lands, the y and -y cancel out but the progress in x remains.

Sports are a preliminary to all the general affairs of society. There are many ways for youngsters to organize an informal game. The ways of the playing field thus seep into the ways of society. Is it only an accident that there are two teams in a sports event and usually two major political parties?

The innings of a baseball game is a mirror image of the back-and-forth of American politics between Republicans and Democrats. The passing of the ball to another player is a mirror image of the stages in a revolution. The balancing on a bicycle mirrors the leaning to the left, leaning to the right and, the center of politics.

Free enterprise is based on competition in the same way as sports. The entire free enterprise system is a game. This game is based on the principles of supply and demand. As with so many other things, this is mirrored in sports. When a young athlete exercises to build up his body, the muscles operate on exactly the same principles as the economic system. The oxygen debt built up in aerobic exercise operates in a way similar to the cash debts in business. Any one who exercises should be a whiz at economics.

Organizing a sports team is done in a way similar to the organization of a corporation, matching the talents and experience of the workers with the job positions. Companies that are trying to gain customers are similar in concept to a team trying to gain possession of the ball. Both involve predictions, reactions and, the search for opportunities, which are done according to similar patterns. Picking which team will win prior to the event works on the same concept as investing and a runner deciding whether to try to get in one more base makes the same kind of decision as an investor deciding whether to make an investment.

My conclusion is that the primal simplicity of everyday sports makes it an inevitably powerful model. The incredible fact that has been largely forgotten is that athletics exists in the same universe as academia. The basic patterns in the two are the same. They are not something "different" but are as the two parallel tracks of a railroad line. Sports should be a very powerful learning tool, far more powerful than it is now. At some level, we must know instinctively that all of this is true.

IDEA # 137; TOP RAIL OF CHAIN LINK FENCE: The top rail of a chain link fence has much more potential usage than it is used for now. For beginners, it could be used as the conduit for a lawn sprinkler, neatly covering the entire yard. Or maybe just for flower planted along the fence. It could be used as a conduit for wiring to power one or more lights, which may or may not be mounted on the fence posts.

IDEA # 138; WARBALL AS ORGANIZED SPORT: There is a gym class game from grade school and junior high school that stuck in my mind. Does anybody remember warball? There are two teams, each standing at opposite walls of the gym. There are usually several balls in play, large balls about half the size of a basketball but not as hard. There is a line across the center of the gym. Players on each side must stay on their side of the line.

That's basically it. You try to take out the people on the other side of the gym by hitting them with the ball between the knees and shoulders. The original purpose of the game was probably, as the name implies, preparation for war. I happen to think that it would make a neat amateur and professional sport.

4

EVERYDAY IMPROVISATIONS

Some chapters in this book are about patterns, like the ones thus far. This chapter is an example or a concept chapter, showing examples of everyday improvisations rather than the patterns in the ideas. Everyday improvisations are the core of new ideas. As they say, 'necessity is the mother of invention'. Look at everyday inconveniences and you will see a myriad of potential improvements. This is the basis of free enterprise.

I would also like to say at this point that I believe that too much respect for the world as it is causes a detriment to progress because it tends to subdue the flow of new ideas. Those cultures putting emphasis on group cohesion have a very low rate of coming up with breakthrough new ideas.

In contrast, it was the Protestant concept of the world as sinful and inefficient and it's think-for-yourself mentality that set off the Industrial Revolution and brought about the modern world.

In the development of new ideas, it also helps to have creative challenges when young. I remember in my late teens and early twenties, I never had enough time to do all the working out that I wanted to. So I got creative with my exercise program, trying to get the maximum benefit from a given amount of time. I did not have enough knowledge or experience to dream of writing books at that time but I did develop the creative mindset.

I also believe that activities like drawing and woodworking utilize the creative mindset. Reading is far more beneficial than watching television if for no other reason than reading requires use of imaginative capacity.

IDEA # 139; ROLL OF PAPER IN PEN: A pen and paper go together. One is not much good without the other. How many times, when it is necessary to write something down can you find one but not the other? There is unused space inside a pen between the case and the refill. This would be an ideal place to put a small roll of paper. The paper could come out through a slot in the case the length of the pen by twisting or could be retrieved by opening up the case of the pen. When one roll was used up, a new roll could be inserted. Paper could conceivably be wrapped around the outside of the pen.

This probably has not come about yet because pens and paper tend to be manufactured by different companies.

IDEA # 140; CURVING TELEPHONE MOUTHPIECE FOR HIGHWAY PHONES: This would be useful for any telephone located in a noisy environment. This idea uses the curvature of the telephone mouthpiece to maximize capture of the speaker's voice while minimizing background noise.

As a wave of sound moves out from it's source, it forms a circular pattern continually increasing in size with increasing distance. Since the speed of sound is roughly the same in all directions, the source of sound is always at the center of the circle as long as no barriers are involved. This means that just after the sound is made and the wave has not yet traveled very far, the circle covered by the wave is very small.

In a small circle or arc, the ratio of curvature is very high. That is, the curvature of a small circle per centimeter or inch of circumference is higher than in a larger circle. The curvature of a larger circle is "flatter" than that of a smaller circle. This means that if we had a microphone with a certain curvature, it would have a "focal point" just as a lens or curved mirror does. Sound with it's source at this focal point would be received by the microphone more efficiently than sound with it's source at any other point.

Why not produce a microphone for telephones with a curvature designed to receive the user's voice? Sound from a more distant source would hit the curved microphone with a flatter wave and so would be out of focus and would not be received as well by the microphone

IDEA # 141; RUBBER BOOT TELESCOPE MOUNT ON CAR WINDOW: Amateur astronomy is a fascinating hobby. One of the problems is finding s suitable place to observe away from city lights and obstructions. I believe that a car would make an excellent mini observatory after the engine has cooled. All that is needed is a rubber boot device to mount the telescope on a partially lowered car window.

The window would be lowered about a third of the way and a right angled mirror device would probably be mounted with the eyepiece to make viewing easier.

IDEA # 142; BUSBOY SQUEEGIE FOR TABLE WITH WAIST LEVEL WASTE CONTAINER: Why not adapt the standard window squeegie to help busboys in clearing tables? This would improve speed and service. A restaurant table is not earning any money when it is dirty. If the busboy has a container at waist level hanging by a neck strap, just below the level of the tables, it would leave both hands free for clearing the table. The squeegie should be of an appropriate size relative to the table, such as half width.

IDEA # 143; WRITER'S COMPUTER APPLICATION: It would be great to have an application for writers. If I had this application, this book would probably be done by now. Suppose a writer could put hundreds or thousands of notes into an application. Then bring up a window to write in while looking over the notes. This would help the writer to begin organizing the mass of notes into 'blocks' of related notes.

These blocks would probably become the chapters or sections of the book. The writer could click on each 'note' and indicate which 'block' it was to be placed in. He could then look the whole thing over and decide to combine, eliminate or, create new blocks. Just by clicking, he could move notes from one block to another. He could divide blocks into sub-blocks and arrange the notes accordingly just by clicking. When notes were clicked on in sequence, they would arrange themselves in that sequence.

When notes were in the right blocks, the author could begin clicking the notes into sequence within the blocks and putting the blocks in order. Notes could easily be pasted into other notes. Finally with the notes from each sub-block visible on the side, the actual writing could be done. It all would seem so easy that books would almost write themselves.

All right, I have done enough daydreaming. It's time for me to get back to writing the usual way.

IDEA # 144; FOLDABLE FOUR-COLOR SQUARE SIGNALLER: This is simply a sizable square a meter or so across. It would be divided into four "panes", each a different color. Each color would be bright and easily visible and would occupy a quadrant of the square.

The square would be used for signaling at a distance. I believe that this would be easier than the old semaphore flags. Each display would consist of one of four possible signals, depending on which side of the square was facing up. A display would be simply raising the square to make it visible to the observer and then lowering it back down. A series of displays would comprise a message.

Messages would have pre-defined meanings. A message to a crane operator at a distance might be to "begin lifting now". Displays could be defined by the colors as read from upper left to lower right, such as yellow, red, blue, green or YRBG. The efficiency of the device could be doubled by placing the four colors in a different sequence on the reverse side of the square. The information capacity could, of course, be multiplied by using more than one square.

IDEA # 145; REVERSIBLE WEIGHT SCALE: Suppose there is a barrel of water for washing. The water is pumped out gradually. You wish to keep track of how much water is in the barrel. The solution is to adapt a weight scale so that it does not show how much the barrel weighs in weight units but shows how much it weighs relative to it's net weight and it's gross weight.

The net (empty) weight could be zero and the gross (full) weight could be one hundred. The scale is reversible in that instead of placing an object on the scale and being told how much it weighs, we define a weight or weight range and the scale tells us how much the object weighs relative to that weight or range.

IDEA # 146; TIMED ENERGY FOODS: There are lots of so-called quick energy foods for athletic events and so on. However, it seems to be a one-size-fits-all mentality. The situations that the athlete or soldier is preparing for may be very different. So much is known about nutrition nowadays that there should be energy foods available for various time spans. The simpler the carbohydrate, the sooner it's energy is available to the body. The more complex the carbohydrate, the longer it takes for it's energy to be released. It is now possible to make half hour energy food, one-hour energy food or, one and a half-hour energy food.

IDEA # 147; LETTERS ON ICONS: A computer mouse is a very handy and ingenious device. But I think that it might not be a bad idea to begin looking toward it's extinction. Why not just put a letter on each icon or each mouse choice and make that selection by simply pressing that key. For example, the icon for internet explorer could be activated by pressing 'I' instead of clicking it with the mouse.

IDEA # 148; IDEA #; SMALL TOWN SYNDROME: I have something good to say about big cities. Consider the real notorious dictators of the world; Stalin, Hitler, Saddam, Khomeini, Napoleon, Pol Pot, Idi Amin and, Mussolini. People have wondered if there is a common thread that could help us to understand them and where they came from.

I have found a common thread in the lives of every notorious dictator that I can think of. They are all from relatively small towns. It would seem that growing up in a city gives a person a more realistic view of people and the world and that a notorious dictator comes on the scene when the product of a small town gains power and applies the views of a somewhat shielded and distorted upbringing to the real world.

IDEA # 149; CONVERSATION CLUBS: America is missing a version of the European sidewalk café. There is not usually anywhere to go to get into a conversation. It would be great to have conversation clubs. Such a club could possibly be associated with a library. There would be a posted topic of conversation on a given day, anything from world events to space exploration.

There would have to be some kind of official to monitor the talk time so that no one could monopolize the conversation. A "stick" could be passed around and only the person holding the stick would be allowed to talk so that more than one person could not talk at a time. It would be possible for a side-conversation to split off and go to another table. Depending on how many people showed up, more than one group could be formed.

This would be a great benefit to society. European sidewalk cafes and pubs have always been think tanks and incubators of new ideas. Most people would be surprised at how many of the ideas that have shaped the world around you have come out of discussions in such cafes. It is time to bring this idea to America.

IDEA # 150; KEYBOARD GRAPHICS: A picture is supposedly worth a thousand words. Wouldn't it be great to be able to draw simple pictures to put in an email or document without a special program? Simple drawings could be made by drawing a line using the mouse or selecting from pre-drawn lines, circles, boxes, etc.

IDEA # 151; MOUSE TYPING COMPUTER PROGRAM: The alternative to eliminating the mouse is eliminating the keyboard. A virtual keyboard would be shown on the screen and letters would be selected by clicking. This would obviously not be easier for the average person. But the computer is a great thing for the physically handicapped. Those unable to use a keyboard could benefit greatly from a specially adapted mouse.

IDEA # 152; DISABLED CAR SIGNAL: One thing that looks rather primitive is the placing of a cloth in the window of a car to indicate that it is disabled. I realize that car manufacturers do not want to reckon with the fact that their cars may actually break down. However, it should be possible to install some simple device, such as a flag that does not use power in a car, to indicate that the car is disabled.

IDEA # 153; MOSQUITO FLUID: I believe that the way to get rid of mosquitoes is by drowning. It is usually stagnant pools of water that breeds mosquitoes. The mosquitoes walk on water by making use of the surface tension of the water to support their weight. Suppose we had a liquid that was lighter than water and had less surface tension. If we poured this liquid onto waters where mosquitoes lived, they would fall through and drown when they tried to walk on the water.

We do actually have a liquid with suitable properties, gasoline. Indeed, gasoline will drown mosquitoes when poured on water. Unfortunately, gasoline has it's drawbacks when used for this purpose. It is expensive and damaging to the environment. Also, Murphy's Law seems to insist that just when gasoline has been poured everywhere, someone will inadvertently drive by and throw a lighted cigarette butt out of the window.

Suppose we could come up with another liquid that would be clear, inexpensive, environmentally safe and, non-flammable that would serve to drown mosquitoes. The liquid should evaporate after a reasonable period of time leaving mosquitoes that tried to walk on it drowned but the environment unaffected.

IDEA # 154; VENTILATE PARKED CAR IN HOT WEATHER WITHOUT WINDOW OPEN: This one is way overdue. Cars build up some unpleasant heat while parked in the sun during the hot weather. Leaving the windows down a bit is unfortunately an invitation to thieves. Most cars do not have sunroofs. Let's have a strategically positioned vent through which rain cannot enter but heat can exit. The vent must be easily to close when the weather is cooler.

IDEA # 155; PHOTOCOPY ONTO EXISTING FORM: A process that would save much work is one that would enable copying with a photocopier onto an existing form. What about taking a mirror image of a form and projecting it onto a blank area of a paper form and then making a standard photocopy. There is so much more than can be done with existing photocopiers with a little imagination. We can start by equipping photocopiers with two bays for paper instead of one. One will be the standard "real copy" bay and the other will be the "virtual bay" which will result in a mirror image which can be projected onto the form in the "real copy" bay. Alternatively with only one bay, a "photo" could be taken of the form in the copier that could then be projected onto future forms placed in the bay. The possibilities seem to be almost endless.

IDEA # 156; PADDED TOOLS: Sometimes, hand tools must be used in high places. Needless to say, a dropped tool can cause a lot of damage. The very expensive mirror of the Palomar Observatory telescope in California was damaged by a dropped wrench. Such tools should be padded and weighted so that the padded part hits first if it falls. In addition, tools used in water can be made to float.

IDEA # 157; METHOD OF DETERMINING AGE OF A NEWSPAPER CLIPPING BY CHEMISTRY: Often, old newspaper clippings are found in which the date is either not present or not readable. Fortunately, newsprint weathers at a steady rate. Why not test newspaper clippings of various known ages and develop a chart of chemical change of newsprint over time. Then when an undated clipping is found, a sample can be taken and tested and a probable date established.

IDEA # 158; COORDINATED AIR RAID: I really do not wish to put forth ideas to make destruction more effective but I suppose that it is better than having a future enemy think of it first. Also, it may help a reader to find a way to use the idea for another application.

Despite the smart bombs today, bombs still explode individually. Each bomb, when it explodes, creates a shock wave that travels through the ground as well as the air. An air raid on a wide area could be made more effective by timing the bombs to explode in such a way that would cause the shock wave of each bomb to interfere constructively with the others. This would make it possible to set up a massive shock wave that would be impossible if the explosions of the bombs were uncoordinated.

IDEA # 159; COFFEE CUBES: I believe that selling coffee in cubes would make the process of coffee making faster and more efficient. In a restaurant, customers ordering coffee typically have no choice in how strong they would like their coffee. If coffee were brewed by adding cubes to hot water instead of by the pot as is done in most restaurants now, customers could choose one, two or, three cubes. One cube could be approximately equal to a teaspoon. This would also be a lot less messy than handling loose coffee.

IDEA # 160; ON HOLD LINES THAT RETURN CALLS: Isn't it great to make a phone call and find yourself on hold? "Your call is very important to us", "just listen to this music for fifteen minutes and the next available operator will be with you".

Why cannot a recorded voice inform you that there is a wait and ask if you would liked to be called back? If so, you press one. Then when an operator becomes free, the return number of the next person in line is automatically dialed.

Why is it not possible that if a phone number is busy, the caller can hang up and the number keeps being redialed until not busy and then the caller's end rings?

I am sure that this is possible now. When you dial a number and it is busy, why can you not just press a button to indicate if you wish to keep trying the phone number? If so, you would simply hang up and then when the phone system was able to connect to that number you would receive a call back and the other party would be on the line.

Since we now have instant messaging, why not combine this with the telephone system so that an instant message can be sent if the telephone on the receiving end is busy and will appear on a screen?

IDEA # 161; SOUND-ABSORBING PILLOWS: Pillows are inevitably made for comfort. However, there is one element of comfort that I feel could be added. Most pillows are not made of material or do not contain material that is designed to absorb sound. In our sleep-starved era when many people sleep during the day in order to work evenings or nights, this would be a big plus.

IDEA # 162; SNAP LOOPS TO REPLACE ROPE OR CHAIN: Why not sell packages of loops made of strong wire or some similar material with a snap on each end that snaps or links together to form a closed loop. The pieces would probably be about a foot long. Whenever a rope or chain is necessary, as many loops as necessary could be linked together to accomplish the task. When additional strength is required, each loop could consist of two or even more links. When the task was completed, the loops could be easily disassembled. This would make ropes and chains largely obsolete.

IDEA # 163; SHOPPING CART CALCULATOR: An ideal place for a built-in calculator would be on a shopping cart. Shoppers can occasionally be seen who bring their own calculators. Scientific functions would not be necessary on the calculator but it should definitely have a function to automatically calculate and add on sales tax.

This can increase the store's revenue because some people probably leave some products on the shelf because they are not sure if they will have enough money. The calculator will show exactly how much the purchases will cost. If a store would really like to make things easier, there could be a magnetic sensor on every cart to automatically read every item placed in the cart and keep a running total, tax included.

IDEA # 164; ANTI-SPIDER WEB COATING: I am sure that there is a market for a wall coating that would discourage or prevent the construction of spider webs. Aside from walls, just about any exposed surface in a room would benefit from such a coating. Research just what it is that joins a spider web to a wall and then develop a chemical coating to counteract the process. The coating could be combined into paints or sold separately. Depending on the consistency of the coating, it will be applied either by brushing or spraying.

IDEA # 165; ASSEMBLED STANDS FOR RUNNER BEANS: Many people that grow some of their own vegetables include runner beans in their gardens. The traditional structure for supporting runner beans resembles a teepee and requires some effort to build. The structure is usually made of wood. I am surprised to have not yet seen a kit of pieces easy to snap or bolt together to support the growing of runner beans.

IDEA # 166; ROLLER FOR LAST BIT OF TOOTHPASTE: An incredibly amount of toothpaste is wasted because it is so difficult to get it out of the tube. Every bathroom should have a simple roller device to get the last toothpaste in the tube. In the long run, it would save quite a bit of money.

IDEA # 167; ELECTRIC CAMERA: A camera, whether using standard film or digital, could be of much more use if it could be operated electrically instead of manually. This would make everyday cameras operable by remote control. The remote control could be accomplished in any way desired, such as a cable or radio remote control. A small electric motor for film winding could also be remote controlled. This would make taking photos of such things as dangerous situations and animals much easier and would also enable the mounting of a camera on the outside of a car.

IDEA # 168; RESTAURANT TABLE SHIMS: Why is it that the majority of restaurant tables seem to be unbalanced? It rests one way until someone puts pressure on the table at the other end and the table shifts that way. If you take a look you will see that all kinds of things get used for table shims, in an attempt to keep tables level. The most common shim is a folded-up piece of cardboard, but that does not always work and looks unprofessional.

Why not just make restaurant table shims that will be unobtrusive and nearly invisible? The shims will be of a wedge shape and most likely made of a rubber-like material.

IDEA # 169; TRAFFIC LIGHT MIRROR: Sometimes, a driver will drive a little too far when stopping at a red light. He will have to duck down and look up to see when the light turns green. For someone with a medical problem, this can be uncomfortable. Why not just place a small discreet mirror at the edge of the dashboard where it meets the windshield. Then the driver could look up and see when the light turned from red to green with out bending down.

IDEA # 170; CASH REGISTER ABLE TO DO MORE THAN ONE TRANS-ACTION: I think that it is about time to have cash registers able to do more than one transaction at once. How many times have you had to wait while someone in front of you in line had to go and get something and you had to wait before being rung up? Or someone had to come and get the order before you could be rung up? There should be cash registers that can put that customer's transaction on hold and go to a B mode and ring up your transaction so you will not have to wait. Ultimately, cash registers will probably work like computers and it will just be a matter of bringing up another window.

IDEA # 171; COFFEE CUP TOP ABLE TO HOLD CREAM AND SUGAR PACKETS SECURELY: This sounds simple but it is yet another way to improve efficiency. When someone takes out a hot drink such as tea or coffee, they usually put the packets of sugar and cream on top of the cup and walk to the table or car before putting the cream and sugar in. Why not make the plastic tops for hot drinks with appropriate indentations to securely hold two cream containers and two or more sugar packets.

IDEA # 172; VERTICAL EYE FOR SUPERMARKET CHECKOUT CON-VEYOR: When you put your items to be checked out on a supermarket checkout conveyor belt, there is an electronic eye that stops the belt when your items get to the cashier. This is so the belt will not keep going forward and cause any bottles and cans to fall down.

The fault of this system is that newspapers and magazines go below and are not picked up by the horizontal eye. On one occasion, I had a magazine go down in the gap above the belt when it got to the cashier. The solution is to place an eye vertically as well as horizontally. This will catch any newspapers or magazines not caught by the horizontal eye and stop the conveyor.

IDEA # 173; DEVICE THAT AUTOMATICALLY TURNS DOWN TV OR STEREO TO PRE-SET VOLUME WHEN PHONE IS PICKED UP: What happens when the phone rings and the television or stereo is playing loudly? Either you must hurry and try to turn it down before answering the phone or try talking with the loud noise in the background.

Why not have a connection between the phone and the stereo and television so that whenever the phone is picked up, either to make a call or to receive one, the stereo or telephone volume is automatically reduced to a pre-set level. When the phone is put back on the hook, the stereo or television volume will go back to it's former level. The connection could be either by wire or infrared.

IDEA # 174; ALTERNATING FLASHES ON EMERGENCY VEHICLE LIGHTS SHOULD INDICATE SPEED: This would add to the safety of emer-gency vehicles. A vehicle speeding through traffic and going through red lights is dangerous. It is often difficult to judge how fast an emergency vehicle is going. Why not set the flashing lights of the vehicle so that the faster the vehicle is going, the faster it's lights will flash? This will be an inexpensive way to increase safety.

IDEA # 175; NEWSPAPER PAGE SET UP AROUND FOLD IN PAGE: Peo-ple read the newspaper at odd times and in places that are not really ideal. It makes it more difficult by having to keep opening the wide pages of the paper. I realize that it is necessary to produce the papers as inexpensively as possible. However, arranging the stories so that the horizontal fold in the newspaper is between stories will be easier to read. The less folding and unfolding of the news-paper that is necessary during the course of reading, the better.

Also, I believe that the newspaper would be much easier to read it stories were just written in one place instead of starting on page one and continuing on another page. This is done so that a potential buyer can see more of the top stories on the front page. However, my observation is that anyone who buys a newspaper tends to buy it anyway regardless of how many top stories are started on the front page.

IDEA # 176; HAND CREAM IN STICK: Many people in many occupations find their hands getting dry during the course of the day. The apparent solution is to keep a container of skin lotion handy. The problem is that such a container is somewhat bulky and is not something that a person can easily put in their pocket. The true solution is to make skin cream into a form like lip balm in a small stick that can be easily carried around. The cream would probably have to be more concentrated than that in the large bottles.

IDEA # 177; HUNDREDS OF MILLIONS OF TREES: Sorry that this is not a new idea. But it is so important that the world needs to be reminded of it. What the world really needs is simply hundreds of millions of trees.

Historically, we have considered trees as something to be cut down as land is cleared in the name of progress. The world needs to get rid of this outmoded idea and do it as soon as possible. The world is undergoing a population explosion while there are certainly fewer trees than there has been in many millions of years.

We need an active, rather than a passive, approach to conservation. Trees literally appear out of thin air. A tree pulls the carbon out of the carbon dioxide in the air and makes it into useful wood. Otherwise, carbon dioxide acts as a so-called "greenhouse gas" and raises the temperature of the planet. The terrible heat wave in Europe in summer 2003 should serve as a dire warning. Trees also, of course, release the oxygen in carbon dioxide back into the air so that the animal kingdom can breathe.

Trees also have a vital role in binding the soil. Trees are the link between the ground and the air. One thing that drove millions of Italian immigrants to the new world was the devastating effects on farmland of deforestation in southern Italy and Sicily in the late Nineteenth Century.

Trees have a strong influence on natural beauty and quality of life. Trees curb erosion and clean the air. The Amazon region has been described as the "lungs" of the planet. My idea is that as much of the planet as possible should be the "lungs". We need government agencies and NGOs, as well as the U.N. to get programs going to plant trees anywhere that they can practically fit in.

IDEA # 178; CAR VACUUM CLEANER: Why does anyone have to go to one of those vacuum cleaners at some gas stations to clean their car out? A car engine creates a powerful vacuum. A car engine could have a function in which it uses engine vacuum to operate a small cleaner attached by a hose to the engine. The engine would, of course, have to be idling when vacuuming was being done. Enough vacuum could possibly be achieved by use of the exhaust manifold rather than by directly tapping the cylinder vacuum.

IDEA # 179; STUDENT'S CALCULATOR TO PRINTER CONNECTION TO SHOW WORK: Suppose there is a physics class in which students are required to show the work on a test or homework assignment. In other words, show just how they came up with their answer step by step. Most science students use scientific calculators in solving test and homework problems.

Why not have a scientific calculator with an extra memory chip to go back over previous work if the memory function is activated. It would even be possible to set up an infrared connection between the calculator and a printer to print out the information that was stored on the calculator memory chip.

IDEA # 180; GLOWPORT FOR MARMALADE, JAM AND, HONEY JARS: I thought of something that would look really neat. When jars of marmalade, jam, honey, etc. are set on shelves in stores, put a small light under each jar. The jars would be set in a semi-darkened cabinet and the lights would be on in the front row of jars. This would make the contents of the jar seem to glow.

IDEA # 181; PHONE THAT TELLS WHEN OUT OF ORDER: I know that pay phones will probably eventually go the way of the dinosaur. Until then, why can't a payphone at least let you know when it is out of order? Since a phone is connected to a source of electricity anyway it should be a simple matter to have a small light, even a LED, on when the phone is operable (or vice versa). How many times have you messed around with a payphone only to find that it is not working? Or went all the way over to a phone only to find that it was a wasted journey?

IDEA # 182; NEWSPAPERS FROM CAR WINDOW LEVEL: It seems as if the design of newspaper boxes has not changed in decades. Today, people drive much more than walk. Newspaper boxes should be designed to accommodate drive-up customers rather than just walk-up. It is really difficult to buy a newspaper from a box when in a car. Payphones have long been designed for drivers, why not newspaper boxes?

IDEA # 183; SOLID PLAYING CARDS FOR THE BEACH: A picnic would be a nice place for a card game. The trouble is that a breeze makes such games precarious. The solution is solid playing cards, probably made of some hard plastic.

IDEA # 184; WALLET-SIZED BAND-AID CARDS: How often could you use a bandage but there is none to be found? Carrying a Band-Aid in the wallet is a solution but it tends to get wrinkled and lose it's effectiveness. A better solution is to have a solid card the size of a credit card in which a segment can be broken off. Part of the segment will peel off to yield a Band-Aid.

People in America tend to come in more races and skin tones than when Band-Aids were first introduced. It is time for a variety of colors.

IDEA # 185; THERMOMETER PAD CALIBRATED FOR CAR HOOD: If a thermometer could be fitted into a pad that could be placed on the hood of a car, it would be useful for determining how long the car has been parked. Such a device could be useful in crime investigations and for private investigators.

There is a natural logarithmic function for heat loss by a hot object in cooler surroundings. The temperature of the air away from the car could be measured separately. The pad would simply measure the temperature of the car hood. A calibration chart could be made of hood temperatures for various makes of car at the instant of stop. With this data and the hood temperature, it could be determined how long the car had been parked. Effectiveness would assume, of course, that the car had been driven long enough to reach the standard operating temperature.

IDEA # 186; REFLECTOR FOR WHEN CAR HEADLIGHT BURNS OUT: Driving with one headlight is dangerous not so much for the driver's visibility as for the visibility of the car to other drivers. A car with only one headlight looks like a motorcycle to other drivers.

A car carries a spare tire, so why should it not carry a spare headlight. It does not have to be an actual headlight, simply a reflector to be placed over the burned-out headlight to make the car visible to other drivers. A simple round mirror attached to a bar perpendicular to the mirror will do. The bar will either be magnetized or will have a magnet to attach it to the car body above the head-light.

IDEA # 187; THE MATHEMATICS NOTEBOOK: When writing a long mathematical statement, it is not as easy to continue the statement on the line below as is done with writing in words. Conventional school notebooks are designed for writing in words, not in mathematics. Why not design a notebook especially for mathematics? It would be used horizontally whereas a word note-book is used vertically. This would make for fewer lines on a page but the lines would be longer. Also, mathematics should have more vertical space on a line than word writing.

IDEA # 188; DRIVER'S SLIPSTREAM INDICATOR: You have heard about how a driver of a car can save fuel in highway driving by driving in the slipstream of a truck in front. What we need is some kind of indicator that fits on the front of the car to tell the driver when he is at the distance of greatest efficiency behind the truck.

The indicator could stick on the windshield and use a tiny windmill that gen-erates a small electric current that moves the needle on a meter. This will show the driver of the car the point behind the truck where the car's impact with the air in front of the car is lowest and thus where the greatest efficiency is located.

IDEA # 189; USEFUL CONTAINER PRINCIPLE: Have you ever noticed how many tasks product containers get put to? This is especially true with coffee cans, jars and shoeboxes. If a manufacturer is looking to gain an edge on his competi-tors, why not consider the container that the product comes in. Make a container that is useful around the house after the product is removed and you will make the product more valuable. Design containers, when possible, not just for show in the store but for usefulness long after the product is gone.

IDEA # 190; CLOCK THAT STOPS WHEN EVENT OCCURS: A device that would be useful in many applications is a clock designed to stop when some event occurs. The clock would most likely have a sensitive switch that would stop the clock when pulled. The switch would be made so that it could be attached to another object by string or wire. The clock could have an internal liquid switch that causes it to stop whenever the clock is moved.

The clock would be small and easy to conceal. It could be attached to a door to tell parents what time their teenager got home on a Saturday night. It could be attached to a door in a sensitive building to show if anyone had entered the building and when. It could be placed in a drawer or filing cabinet to show if it was opened and when.

IDEA # 191; COMPUTER PROGRAM TO DISPLAY ALL POSSIBLE ALGEBRAIC MANIPULATIONS OF AN EQUATION: Suppose you had a complex algebraic equation. Something like: AB/C = (DEF) squared x G/H x IJ/K. Suppose you wanted to redefine the equation in terms of G. In other words find G = ? It would take quite a bit of work to isolate G from the rest of the equation.

This is just the kind of thing that computers were made for. Yet, I have never seen a program that will isolate a variable and rearrange an algebraic equation. Such a program will be useful in solving a wide variety of problems.

IDEA # 192; NATIONALITY DICTIONARY: I have yet to see a nationality dictionary. It would be an ideal addition to a standard dictionary or a world atlas or on the wall of a classroom. It would simply show the associative and individual aspects of the world's nations. Anything associated with Poland is Polish and a native is known as a Pole. Aside from nations, many cities and regions could also have entries in the dictionary.

IDEA # 193; TRIANGULAR GLASS BAR TO PHOTOCOPY BOOK: Have you ever noticed that when photocopying a page in a book, that books obviously were not designed for ease of photocopying? In the middle of the book, the two pages form a valley and the information in this valley is distorted during photocopying.

What we need is long, triangular glass bars of a few different sizes for different sized books. The bar would fit into the valley of the book in the photocopier and would use refraction to compensate for the resulting distortion. Alternatively, one side of the glass bar could be mirrored and compensate for the distortion by reflection. The bars would be kept near photocopiers anywhere that books were likely to be photocopied, in copy shops and college and high school libraries. The page of a book could then be photocopied without the usual distortion.

IDEA # 194; ELECTRIC WINDOW SCREEN TO TAKE CHILL OFF INCOMING COLD AIR: Suppose you want some fresh air through the window but it is just a little too chilly. Why not put some electric nickel-chromium wires in the screen to give the incoming air a little warmth? It will not be wasting heat by sending it out the window if the air is coming in rather then going out. Not all of the wire in the screen needs to be heating elements. The electric screen can be powered by plugging it into a wall outlet.

IDEA # 195; HAND SIGNALS BOOK KIT: There are many environments in which hand signals could be useful. Industry in which there is noisy machinery and frequent communication must be accomplished over some distance is the first instance to come to mind. It could also include situations in which close communication was necessary but the sound of speech is not desirable ranging from soldiers near an enemy to law officers conducting a raid or other operation.

There is, of course, standard sign language for the hearing impaired but it is usually only a few situation-specific communications that are necessary and sign language was designed for close-range communications, not for a distance of ten meters or more.

The solution is to classify and photograph common positions that the human arms can be held in. Then photograph a model in such positions. The organization intending to use this communication method can then assign meanings to each position. The two arms held straight up over the head could mean one thing. One arm held up can mean another thing. The arms crossed at the middle of the forearms can mean another thing. It is necessary only that the sender and the receiver both understand the meanings of the signals.

IDEA # 196; TURTLENECK SHROUD TO NIP COLD IN THE BUD: Sometimes, when you feel a cold coming on with that scratchy feeling in the throat, it can often be "nipped in the bud" before it really takes hold. A cold is so called because cold temperatures weaken the body's defenses against the germs.

When the cold is first felt as the scratchy feeling in the throat, it can often be nipped in the bud by putting a towel or scarf around the neck overnight and keeping the throat warm in particular. The problem is that the towel often comes off during sleep or is difficult to keep on while awake. The solution is a specially designed band made of some material like wool to keep in place on the throat and so keep it warm enough to ward off the cold germs.

IDEA # 197; SOAP BAR THAT DOES NOT BREAK: I am surprised that no one has yet come up with this. So much soap is wasted because the bar tends to break into pieces when it is almost used up. I believe that this can be accomplished by making the bar less solid in consistency and more like clay.

IDEA # 198; CHECK TIME WITHOUT LOOKING AT WATCH: It must be possible to make a watch from which the time can be felt by running the fingers over it. In some situations such as meetings, it may not look good to be seen looking at your watch. In particular, George Bush Sr. lost a lot of points with voters in his 1992 debate with Bill Clinton by looking at his watch. It came across as if the Republicans could not match Bill Clinton and Bush was worried about scoring some points before his time was up.

IDEA # 199; KIT TO WRITE ON CONCRETE: Have you noticed how people like to write things in wet concrete? Names, initials and, dates are to be found written in concrete sidewalks all over. The writing is usually done with a stick. Why not make a kit especially for writing in wet concrete before it sets?

When a home is being built, the family can decide on a phrase out of the Bible to be written into the concrete. Special events can be memorialized in concrete. A kit may be available from a rental store with a solid background and raised letters that can be mounted on the backing and placed face down in the wet concrete.

IDEA # 200; WEIGHING PET: Pets such as dogs and cats have bodyweight issues too. Have you ever tried to weigh a pet? It usually means holding the pet while trying to look down at the scale. If you have one of those balance scales like the ones in doctor's offices, you must try to move the weights while holding the pet. The only sure way to do it correctly is to have someone help you.

It would be easier with a basket specially made to fit on the scale, probably with two supports underneath the basket that fit neatly on the scale. The basket would be of a known weight such as exactly five or ten pounds. The pet would be placed in the basket, which would then be weighed. For larger dogs, a flatter platform could be used instead of a basket. Possibly the sides of the basket could be let down to for a flat platform for the larger dogs.

IDEA # 201; SHAPE-CODED LABELS FOR DARKROOM BOTTLES: I realize that digital cameras are gradually pushing conventional darkroom photography aside. But it will be some time before photography is entirely digital. A darkroom is so called because it is dark in there. There is a reddish light in most darkrooms that will affect developing film but it is still much more difficult to see than it is out in the light.

There are chemicals that are used in the darkroom first to develop the negatives and then to develop the prints. I think that this would be much less likely to result in use of the wrong chemical at the wrong time if darkroom bottles had shape-coded labels. The labels would have a raised surface. A square label, a circular label and, a triangular label could indicate the different chemicals.

IDEA # 202; DRILLING OR NAILING REAR SUPPORT: A problem arises when one person is building a fence or must nail or drill into a post in the ground. Unless the post is very solid and secure, the action of nailing or drilling puts pressure on the post that can force it to lean to the other side. Usually, the handyman has to awkwardly hold the post from behind with one hand while nailing or drilling with the other hand or else have a partner lean his shoulder against the post.

   The solution is a belt that the handyman will wear, possibly across the back and one shoulder. The belt will have a hook that clasps the post from behind. The handyman will counteract the force from the nailing or drilling on the post by leaning back slightly so that it puts a counter pressure on the post. This will be much less awkward and more effective as well as leaving the handyman with both hands free.

IDEA # 203; RUNNER'S LAP COUNTER: When a runner is running around a standard oval track, it is a distraction to keep track of how many laps have been done. It is easier to just do it by time but that does not give as accurate a representation of how far has been run. Why not just combine a compass and a counter to measure how many times the runner has gone around a circuit of the compass directions? The device will be worn by the runner and will be modified so that it will not be affected by vertical motion such as bouncing.

IDEA # 204; EXERCISE REP COUNTER FOR CALISTHENICS: Some people like to do high numbers of pushups or sit-ups while exercising. As with running, it is a distraction to keep count unless there is an exercise partner present. A counter that measured vertical change of direction would be of great help. The counter would be small and easy to clip onto the waist or the sleeve and, of course, able to be set back to zero at any time.

IDEA # 205; FILLABLE TRAVEL SHOWER: When camping or travelling, it would be great to be able to take a shower anywhere. Why not make a container that can be filled with enough water to take a shower and which can be hung on any support. The water could be heated first and a simple plastic privacy curtain could hang from the shower.

IDEA # 206; PUSH POLE FOR SAMPLING UNDERGROUND: To get a quick sample of what is to be found underground at a certain depth we can use a push pole consisting of one tube within another.

The pole is driven or twisted into the ground to the desired depth. Then, a separate handle is attached to the end of the pole and the inside tube is driven into the ground a little further, during which time the outside tube remains stationary. The inside tube is then pulled back into the outside tube and the outside tube is pulled out of the ground. A sample of the ground at that level will be in the end of the inside tube. This device will be especially useful in war zones to find such things as mass graves.

IDEA # 207; WIND POWER WITHOUT MOVING PARTS: If wind power is ever going to really take off like it could we need one simple thing: a way to harness the power without moving blades. There will hopefully come a time when those rotating vanes will seem archaic.

Why not build a rectangular box with an opening along one side to admit the wind? The moving air will pass through a screen similar to a household window screen. The screen will function like a grid in a vacuum tube. It will be charged by a source of voltage and so will induce a charge on the atoms in the incoming air. The charged air will then move across or through another screen or surface and, since the air is charged, it will induce an electrical current on the screen or surface.

This will generate electricity with no moving parts except the air itself. I believe that this should be our goal with regard to wind power. Plainly and simply, harnessing of wind power will remain limited as long as those huge towers with rotating blades are required.

IDEA # 208; EASY-OPEN JAR: Many people, seniors especially, have a lot of trouble with products that are difficult to open. The solution for jars is to make the jar top of two different metal alloys. The top should be made of an alloy that expands considerably with increasing heat. The sides of the top would be made of an alloy that expands little when heat is applied. That would make it so that anyone having difficulty opening the jar would only have to run the top under hot water. The top part would expand, loosening the contact between the sides and the glass. The sides of the top would not expand enough to press hard against the glass threads of the jar and so make opening difficult.

IDEA # 209; AUTOMATICALLY CLOSING CAR FUEL PORT: It is about time for the door to the fuel port on a car to close automatically if accidentally left open. It is not really a big deal but it looks kind of dumb to be driving with the fuel port left open. The best solution would probably be to have a port door that has two pieces of metal in contact holding it open but that would be easily jarred loose if the car began moving. Once the two pieces of metal were jarred loose, the port door would close automatically.

IDEA # 210; INCLINABLE BED: I believe that this would be a really great idea. It would consist of two support bars that would be placed at each end of a bed so that the four posts of the bed were resting on the support bars, two at the head of the bed and two at the foot. Each support bar would consist of two strips and a scissors jack. There would be a control box that could raise and lower the support bars.

This would mean that either the head or foot of the bed could be raised or lowered. This would make possible the inclination of the bed. Whenever one has soreness, sunburn or just trouble sleeping, the bed can be inclined to a variety of possible positions to find more comfort.

IDEA # 211; ADJUSTABLE AERODYNAMIC DISC: The aerodynamic disc is otherwise known as a Frisbee. A popular toy and educational in getting future aircraft designers used to aerodynamics. But it can do even more. Why not make the central part of the disk consist of angled, adjustable fins that can be adjusted by a knob in the center of the disc?

The fins can be set flat so that it looks and flies like a conventional Frisbee. Or, the fins can be angled so that the air that contacts the fins when the disc is spinning is diverted downward through the middle of the disc. This would act as a "lever", trading horizontal motion (distance) for vertical motion (height). Instead of going far, the disc would climb high in the air and then come back down. The fins could be adjusted to a number of positions to create a variety of effects.

5

DIFFERENT SCALE

I want to emphasize the difference between an idea and a product. A product is a manifestation of an idea. The patterns of the ideas are very important here. If we can think in patterns, instead of in the given or the tangible product, we open up a whole new frame of reference for creativity. If you explore the patterns of ideas, you will see that many apparently unrelated products share the same idea patterns. An idea that produced one product may produce another product on a completely different level. A good example is the fact that the idea for the lines on a TV screen came from the rows in a potato field.

In this book, I am purposely separating ideas in the same category such as ideas for cars, ideas for mathematics, etc. because I want to throw a light on the patterns in new ideas rather than just listing the ideas by category.

IDEA # 212; CABLE ACROSS DRIVEWAY RECOGNIZING CHASSIS WIDTH: Security experts are always looking for ways to identify people from fingerprints to iris scans, but in these days of car bombings, what about cars? If a car is seen, the model can be identified and the license plate read.

But what about when it is not seen? One possible way to identify cars that pass a certain way is a cable to measure and record chassis width. The measuring cable could consist of parallel wires pressed into contact by the weight of a car. From each side, a source of current could be provided and the resistance recorded. This would indicate how far from each side the outside of the tires passed.

Thus it could be automatically recorded how wide the chassis must be. While is true that a number of cars will have the same chassis width, this would be a simple and inexpensive device that could add to other evidence or at least provide something to go on.

IDEA # 213; CAPACITOR ROOM FIRE EXTINGUISHER: This is a concept that needs testing. Suppose that there was a sensitive room where a fire simply could not be allowed to break out but where water or foam could not be used either, for example in a spacecraft or a nuclear reactor control room.

I wonder if electrons could be used for fire suppression if a fire ever did break out.

Burning consists of atoms combining with oxygen. This combining happens because both the oxygen and the flammable material have openings in the outer layers of electrons in their atoms. The combining requires then energy of heat to get started but it also releases heat and so the burning is self-sustaining.

Suppose that in the event of fire, the room was flooded with electrons. Would that put out the fire? The easiest way to do this would be to build the entire room as a large electric capacitor. Two parallel walls would consist of metal that would be connectable to a large power supply by a switch in the event of fire. The room would store electrons just as the small capacitors in radio or television sets do. Maybe enough electron density in the room would fill outer shells of electrons in the oxygen and material atoms temporarily and prevent the atoms from combining and thus, burning. As soon as the power was disconnected, the room could be drained of excess electrons by a grounding cable.

IDEA # 214; HEARING AID AS MILITARY WEAPON: Soldiers use binoculars and night vision goggles to see more on the battlefield. Why not adapt conventional hearing aids to hear more, particularly in a chosen direction. Maybe a soldier can hear enemy movements. The device can be modified to block sound above a certain level that the soldier would hear without the device anyway. Also, it must be set so that it will amplify the sound of an exploding shell and destroy the soldier's hearing.

I also believe that the manual sign language used by the hearing impaired would be useful for special forces. Suppose two soldiers on a mission wished to communicate when the enemy was nearby but did not want to risk giving their position away?

IDEA # 215; AUTOMOBILE ELECTROMAGNETIC SIGNATURE: Each car must give off a different frequency band of electromagnetic radiation when in operation. This would come primarily from the action of the spark plugs but would be affected by the parts used in the car and also wear on the car. This would be a powerful and discreet way to sense and then to identify a car. With a receiver and directional antenna, it could be used to measure the speed of a car without using radar that is detectable with a radar detector. Without webcams, cars can be counted by reception of the frequency band.

IDEA # 216; POP REAR ENGINE OR TRUNK LID AS EMERGENCY BRAKE ON CAR: Have you ever watched a fast aircraft land on a runway and then pop a parachute in order to lose momentum. Why not apply the same concept to cars? If the brakes fail on a car going fast, stopping could be hastened by popping the trunk or rear engine lid. The lid would have to be supported by a hydraulic jack. This would probably be reserved for cars designed for high speed.

IDEA # 217; THE QUIETER CAR: The loudspeaker of a stereo is designed so that sounds from the front and back either reinforce each other or cancel out. Why do we not put the same acoustic care into car engines that we do into stereo speakers. A car engine compartment is a box just as a stereo speaker is. Surely some of the same principles would apply.

Much of the noise of the car comes from the muffler. I believe that study of the acoustic patterns of any make of car will yield the possibility of placement of the components that will cancel out quite a bit of the noise. Possibly, noise reflector panels could be incorporated into the car to hasten this noise cancellation.

IDEA # 218; NOISE CANCELLING OF JET ENGINES: Jet engines are noisy. There is no way around that. But the noise seems to be somewhat of a monotone, fairly close to a single frequency. How about doing some testing and research to come up with an engine placement plan that would cause the noise from the plane's engines to largely cancel out. Nothing like this is going to make a jet plane really quiet. But, I am convinced that they can be a lot quieter than they are now.

IDEA # 219; TOY CARS WITH PURCHASE: Why not make toy cars into exact replicas of real cars. Whenever someone buys a car from a dealer, they should get at least one toy copy of their car. That way, people could keep souvenirs of all the cars that they had ever owned.

IDEA # 220; SACRIFICIAL ANODE FOR AUTOMOBILE: Ships have what is called a "sacrificial anode". It is a metal that for electrical reasons rusts more easily than the metal from which the ship is made. The metal in the sacrificial anode, usually magnesium, rusts away but does so in place of the structural metal of the ship. Sacrificial anodes are used in ship structures and rudders and also in pipelines. Use of a sacrificial anode works even better than covering the metal with plastic because if steel was covered by plastic and then a small hole developed in the plastic, all the rusting would be concentrated at that one point.

If it can be done for ships, why not for cars? Chunks of magnesium could be joined to the car parts in strategic locations and replaced periodically.

IDEA # 221; AUTOMATIC MONITOR FOR CAR EXHAUST GASES: A car should automatically monitor the exhaust gases for chemical composition. A relatively small sensor would reveal much about what is going on inside the engine. It could even be used to control the fuel-air mix for maximum efficiency.

IDEA # 222; CALCULATOR WITH MEMORY CHIP TO RECALL PREVIOUS OPERATIONS: This is another application of a memory chip in a calculator. Suppose a student wants to go back to a calculation that he/she has forgotten. Web browsers have a history to go back to, why not calculators.

IDEA # 223; SMART SMOKE DETECTORS: Since we live in an era in which all kinds of gadgets are going smart, let's not forget about smoke detectors. A smoke detector uses a small nuclide to detect the presence of smoke. It should act more like the black box on an airplane and analyze smoke also.

If we take smoke detectors up another level, they could even analyze heat levels and record the time. If the detector was made fireproof, it could assist greatly in determining the cause or course of the fire, instead of just warning of the presence of smoke.

IDEA # 224; CHAIR RECORDING SERIES AND WEIGHT OF THOSE SEATED: In these days there are sensitive buildings in which is may be useful to keep track of everyone who enters the premises. For best results this should be done with a variety of methods as a single method, such as a camera, may be fooled or otherwise prove unsatisfactory.

How about chairs made to record the weight of everyone who sits down? Premeasure the proportion of body weight that goes into the chair, as opposed to through the feet into the floor. The chair would have a discreet spring device in each leg that would be added to give the weight put into the chair. A chair could also be placed on a pad on the floor to accomplish the measurement, although this would be more easily spotted. This could be attached to an electrical resistance that could either convey the data by wire or record the data within the chair for later retrieval.

IDEA # 225; PYTHAGOREAN THEOREM IN THREE DIMENSIONS: Has anyone ever noticed that the old Pythagorean Theorem works in three dimensions as well as two? I do not know if this is known already but if it is, I have never heard of it.

The old theorem is used to calculate the length of the diagonal of a right angle when the lengths of the sides are known. The formula is; C squared is equal to A squared + B squared. Where C is the unknown length of the diagonal while A and B are the known lengths of the sides. This is a two-dimensional formula.

The theorem works also in three dimensions. The formula in this case is; D squared is equal to A squared + B squared + C squared. This would give you the three-dimensional diagonal of a room from the bottom corner of one diagonal to the top corner of the other diagonal, for example.

6

NEVER THOUGHT OF

This is another chapter of examples rather than patterns. In our search to develop a better world, before we look for a way to do something we must first think of doing it. In this case, it is not a question of 'can we do it'? But that no one has ever thought of doing it. All around you right now there are things so obvious that we do not notice them. Things we could do easily with existing technology except that no one has ever thought of it.

IDEA # 226; SULFURIC ACID AND WATER PORTABLE STOVE: Combining sulfuric acid and water gives off a tremendous amount of heat. Both are compact and easy to transport. Why not have a portable stove or even a heater that operates by combining the two?

Pre-testing would be necessary to establish the amount of acid to go with a given amount of water to reach a certain temperature for a certain time. Insulation would be necessary, both for the heat and the corrosive effects of sulfuric acid. There would have to be some way of relieving the resulting pressure. Inlets would be needed for both the acid and the water into one or more reaction chambers. The sulfuric acid could possibly be in tablet or capsule form. There would be a cooking or heating surface upon which the heat would be concentrated. Finally, there would have to be a method of draining and cleaning after use.

IDEA # 227; CAR LIGHT SOUND OPERA: Suppose car headlights could be made to turn on and off one at a time. One or more cars could be parked near the stage of an outdoor concert and the headlights could act as flashing strobe lights, the lights being timed with the beat of the music. This would save money on expensive equipment and the musicians would be driving their cars to the concert site anyway.

IDEA # 228; CLASSIFY HUMAN WAYS OF WALKING: This would be a powerful tool in identification in these days of cameras everywhere. On a number of occasions, I have recognized a person from a distance by the way that they walk. I am certain that this could be classified and described mathematically. This is not the exactly same thing as recognition of a person's foot patterns while walking on a floor, which was another idea.

IDEA # 229; ADAPTATIONS FOR THE HOME: Neon signs were to be seen in businesses everywhere at one time. What about a little neon sign with the address for residential use? It would be easily visible at night, when home addresses are difficult to see. Such a sign would not use much electricity and would also serve as a nightlight.

How about a chapel in the home? Why not set one room aside as a home chapel? There is no reason that stained glass windows cannot be made to replace conventional home windows.

IDEA # 230; SNOWDRIFT ART: Here is an art form that apparently no one has yet thought of. The placing of obstacles not as an art form in themselves but to cause drifting snow to create a work of art. I have noticed the artful swirls of snow around objects of various shapes and I think that it would even be possible to set up obstacles in a way that would create drifts in the snow no matter which way the wind was blowing.

IDEA # 231; FACTORY ART: Now that the industrial age is winding down there is nostalgia for those blue-collar union days and those decent factory jobs. I think it is time for an artist to make a subject of factories, whether dormant or still in operation. A factory building with it's lines and angles, pipes and, smokestacks is practically made for artists.

In a couple of decades, old-style factories have gone from monsters bent on destroying the environment to an endangered species. I think people would be comfortable embracing factory art now. For many, a painting of a factory on the wall would bring back memories of good days. Even the word "factory" is starting to sound a little quaint to me.

IDEA # 232; DAYS WITH NAMES: I wonder also why no one has ever named the days of the year. We have named the months and the days of the week but for the dates we have just numbers. If an organization can come up with 366 entities that they would like to give honor to, why not naming a day of the year for each one? There is easily 366 significant characters and places in the Bible. Christmas Day would logically be Jesus' Day.

IDEA # 233: POWER AT THE CENTER: In politics, there are those who believe passionately that the right is best and those who feel just as strongly for the left. I believe that as time goes on, it will become more apparent that the real power is at the center. This means that at the time of this writing in the George W. Bush administration, the United States of America would be much better off by moving considerably to the left. In fact, I feel that America really needs a strong socialist party to raise the issues that improve the quality of life.

The capitalist model was developed a long time ago when social benefits were not necessary because there was plenty of land for the taking. The men who founded the United States went to great lengths to guard against concentration of power. Capitalists, however, practice politics that permit far excessive concentration of wealth for the good of the society.

The Republican Party has somehow convinced tens of millions of Americans that they are the party of both God and patriotism. Americans still fawn over the robber barons of a century ago and marvel at how much money they accumulated, forgetting that they amassed so much money by paying workers a dollar a day, if that.

As a former devout Republican sympathizer, I can tell you that the Republican Party of today is the spiritual descendent of slavery and robber barons. Capitalism may be a productive system but it allows a myriad of evils to flourish. We tend to forget what capitalism really is.

At one time, America used to have what is known as "company stores". A large company used to own the entire town and workers would not be paid in actual money but in "scrip" that was issued by the company and could only be used at the company store. The company could effectively control every aspect of the worker's lives. That was not exactly the freedom that the country's founders had in mind.

I recall reading a story once, about a girl of about eleven or twelve years old. Her family had to move from a farm to work in a factory near Boston because earning a living at farming was proving impossible. The girl was put to work on a machine in a factory building. Being unfamiliar with machinery, one day the girl cut her hand off in the machine. Her bosses simply bandaged her up after stopping the bleeding and twenty minutes later, she was back working the machine again. She was told that if she wanted her family to be able to eat, she would have to manage working with one hand.

I do not have documentation of the story but it is known that similar incidents were common in the late Nineteenth Century. This is what unbridled capitalism is.

One of capitalism's featured moments was the massive stock market crash of 1929. Incredibly, the root cause of the crash was because of low wages. Warehouses were full of goods at the time of the crash but worker's wages were too low to be able to afford the goods. Factories began cutting back, meaning that workers had even less money, and it spiraled into a crash. This folly of capitalism is what really got communism going. The resulting global depression really devastated Germany and made possible the rise of Hitler.

Today, I believe that America's to the right political system grossly over-rewards success and permits destructive levels of wealth accumulation and causes a myriad of social problems. The American people as a whole allow this to continue by adoring those who get rich and marveling over their accumulation of wealth. There are now (2003) tens of millions of Americans scraping by paycheck to paycheck and tens of thousands actually working full time while living in homeless shelters.

Capitalism is productive but it often makes use of advertising to create artificial needs. As someone once said "The purpose of advertising is to make us think we must have something that, in fact, we really do not need". In order to make money, capitalists commonly appeal to the lowest and cheapest impulses of people. The strategic placing of candy in supermarket checkout aisles so that children will pester their parents for some while standing in line, never mind the effect it will have on the children's health. Every advertiser knows the maxim that "sex sells", people will spend money for products if they can be convinced that it will make them more appealing to the opposite sex.

Most people would never believe the amount of science that goes into the layout of a modern supermarket to try to get shoppers to spend more money than they otherwise would. Beware of products placed at eye level and those on end caps of the aisles because that is where the poorest deals are usually placed because that is the spots that are most in the shoppers' direct lines of vision.

Have you noticed that in those countries that are devoutly capitalist, prices almost always seem to end in .99? Pay five dollars for that product? No, that is too much I do not want it. But look at that similar product over there, it is not five dollars, it is only $4.99. That sounds like less, I'll take it. Making it one penny less makes me feel like I am getting a good deal because I am not really paying five dollars. This is the crudest of capitalist salesmanship intended to separate people from their money.

Capitalism is productive all right, operating by the law of supply and demand. But so much of it relies on the less than wholesome demands. Endless hours of mindless television, violent and destructive movies and, truckloads of harmful junk food are prominent among the fruits of capitalism. Tabloid television and periodicals make use of what should be people's private lives to provide cheap entertainment.

One factor in evaluating capitalism is what I will call "The Fluff Factor". People in a capitalist society may work very hard and that is certainly commendable. The question is: work hard at what? Progress is not all genuine, there is also artificial progress.

Artificial progress is that which will show up on a company's balance sheet and create jobs but which will do no lasting good for the society. Naturally, any society wants to make progress and wants as much of that progress as possible to be genuine as opposed to artificial.

One of the problems with capitalism is that it produces a tremendous amount of fluff, or artificial progress. This seems to be progress-jobs and products, but is unnecessary or will add no lasting value to the society.

A capitalist society may have a lot of construction going on but much of it is because the shifting nature of capitalism has made other commercial buildings prematurely obsolete. In Niagara Falls, NY, the town where I live, when I was a child and just landed here, the Pine Plaza was where things were happening and seemed to be the focal point of the LaSalle section of town. Then, they opened the Summit Park Mall and the center of gravity left the Pine Plaza, even though the buildings of the plaza were still in good shape and there was no apparent reason that it should not remain the center of things.

However, capitalist fate was not kind to the Summit Park Mall either. The mall building was built to be good for at least seventy or eighty years. But barely ten years after it opened, another player came on the scene. The Prime Outlets Mall pulled the economic center of gravity away from the Summit Park Mall and left it virtually abandoned, with less traffic now than the original Pine Plaza that it supplanted.

Few people really stop to think how wasteful this is. The constant shifting of a capitalist society means that so much work might go into a building that should be good for close to a century, but market forces will make it effectively obsolete after little more than a decade. This incessant construction is an example of fluff or, artificial progress.

Another example of fluff is the millions of salespeople in every capitalist society. Not that some salespeople are not necessary but when thousands of salespeople are walking around the business districts of America getting business owners to sign up with one phone company and thousands more will go through the same business districts a couple of days later and try to get them to sign with another phone company, it is very wasteful. This is expending a tremendous amount of effort for ends that will produce absolutely no lasting benefit for the society except to switch from one company to another.

A third example of fluff is the legal industry. Capitalist societies seem to be invariably highly litigious. Some lawyers are always necessary but by some accounts, America has more lawyers than the rest of the world combined. This would be fine except that the legal industry produces nothing of lasting benefit to the society. The ideal society has as little litigation as possible.

Possibly the most prominent example of fluff is the automobile. Cars are wonderful things and everyone should have the right to own one except that, at least in North America, cars have been made into necessities rather than luxuries. Society has been set up so that it is practically mandatory to own a vehicle. This is the greatest example of the creation of artificial need, and therefore fluff, that the world has ever seen.

Yet another example of fluff is the law enforcement and prison industries. Incarceration is possibly America's greatest growth industry. These industries have been necessary for all nations. But it is the coldness and meanness combined with the gross inequalities of capitalist societies that have proven to create breeding grounds of crime. This makes the very large number of police officers and prison guards necessary. Therefore, the police and prison industries in high-crime capitalist countries must be considered at least as semi-fluff.

My conclusion is that a capitalist economy too free of government controls will be generally productive but will not only produce destructive disparities of wealth but also tremendous amounts of fluff, or artificial progress, that create jobs but add nothing of permanent benefit to the society. Socialism is defined as modified capitalism and is needed to minimize the waste and social evils that come with capitalism.

Capitalism has a great advantage over a controlled economy, that of incentive. People will naturally work harder for themselves than they will for someone else. The old Communist theory that man is a generous creature that loves to work and to share his wealth has proven to be nonsense. Capitalism is productive because it realizes that what people really care about is "looking out for number one", which is me.

Capitalism's basis is the idea of meritocracy, that people should be rewarded for hard work and industriousness and I for one certainly have no argument with that.

The problem is that capitalism is an over-meritocracy. It allows a lion's share of the wealth to go to those who happen to get the advantage. It allows the people with the advantages to effectively make the decisions. It allows those who get the jump on the others to set everything up to suit themselves.

In case anyone has forgotten, this is an economic version of the kind of dictatorial concentration of power that the United States of America was set up to correct.

Imagine two men, Bill and Bob. Bill is very ambitious and was always studying as a youth. Bob's thing is watching sports and going to the beach. Bob got a job in a warehouse while Bill eventually became the CEO of a large company.

I am all in favor of rewarding Bill for his effort and success and I believe that his pay should be four or five times as much as Bob makes. If Bill was in charge of a really large company, he should earn even ten or twelve times what Bob earns. If we want people to work hard and study then we must have a system that rewards hard work and study.

The problem is that in America today, Bill would be earning at least four hundred times what Bob earns. I believe that this is a very destructive and untenable situation and it is obviously a result of those at the top setting things up to suit themselves. For the good of society, we have got to turn this around. This is gross over-meritocracy. There is only room for so many at the top. We must remember that a system based on the harnessing of greed also risks being distorted by greed.

Aside from fluff, capitalism brings about a number of destructive spirals. One of these is the advantage spiral. Since Bob earns so much less than Bill, it sets up an advantage spiral that is very difficult to break. Bill's children and Bob's children will be born theoretically equal. But Bill has so much money that his children can be sent to the best college with tuition paid and unlike Bob's children at the local community college, will not have to work during college. No straight-thinking person would say that Bob's children have anything like the same chance of success as Bill's children.

Put simply, one of the destructive effects of great wealth disparity is the creation of an advantage spiral and a parallel disadvantage spiral that is very difficult to overcome and contradicts the American axiom that "All men are created equal".

This does not mean that all children must be born into equal circumstances. Forced equality destroys incentive. But it does mean that while one family may earn ten times as much as another, four hundred or more times as much is very destructive. If money is power than those with power will naturally try to set things up to suit themselves, but this is plutocracy and not democracy. If those at the top make the rules, then it can be expected that they will be overpaid and everyone else underpaid.

What about the effects of capitalism on the culture that Republicans and proponents of capitalism do not seem to mention? I do not claim that non-capitalist societies do not have any problems because they do. But capitalism always creates a cold, harsh and, grating environment. The society has a cold and nasty edge to it. The high stress levels produce warped people. The gross income disparities breed crime and the shallowness and emptiness of a materialistic culture breed drug abuse. Bombardment by continued sales efforts forces people to be cold and I believe that cliques and alienation in high school is the result of excessive corporate influence on society. Capitalists have stolen Christmas and made it into a celebration of the Gospel of Materialism.

The world today is not like the frontier days when a man could head west and stake out his own territory. The system itself is too fickle and precarious. A minimum standard of living should be guaranteed to all. A social net is a necessity. All the other advanced countries have such a safety net.

Actually we do have a social net already, it is called prison. I am certain that there are many people who secretly want to go to prison, either consciously or unconsciously simply because it provides a place to sleep and three meals a day. But having prison as our social net is very expensive both financially and culturally. Political conservatives do not tell you that.

Possibly the most destructive effect on the culture of excessive concentration of wealth is the simple callousness that it brings about. How are people with enough money supposed to sleep comfortably at night knowing that tens of millions of fellow Americans are going to bed hungry, or on a cot in a homeless shelter, or packed into a slum with minimal chance of ever escaping. It inevitably forces people to grow a shell, to become callous. It kind of puts a dent into what America is supposed to be about.

I am not saying that we should just "bail them out", that would have a very destructive effect on incentive. But we must be sure we have a system in which everyone can "bail themselves out" and a minimum standard of living guaranteed for everyone.

I believe that an example of how incredibly callous we have become was to be seen a few days before the 9-11 attacks. A distraught woman was preparing to jump off a bridge in Washington State. Carloads of guys were driving by yelling at her to jump. According to some reports, a local radio station actually played the song "Jump" by Van Halen for them to blast on their radios. It was a pathetic display of what kind of callous people we are becoming that was seen across the world on television. We have got to do something to turn this around.

Being in a capitalist society, we tend to think in terms of money. This causes us to forget that there is a very big difference between wealth and quality of life. Socialism concentrates on quality of life whereas capitalism concentrates on wealth. I define socialism as the middle ground in the political spectrum. Capitalism is on the right and communism is on the left of socialism. Contrary to what some people believe, socialism is not the same as communism and is, in fact, as different from communism as it is from capitalism.

Have you ever wondered about the patterns of immigration to the United States? A hundred years or so ago, there were millions upon millions of immigrants from Europe landing at Ellis Island. Today, immigration to America involves very few people from those same countries. A few Europeans still move to America but it is mostly for business or marriage or just for the adventure of it.

What has made this big difference is the implementation of socialism in Europe. It is now a very comfortable place to live and few see any reason to cross the ocean. Those in America who criticize socialism tend to ignore this. They cannot say that socialism does not work because obviously it does. The pattern of European immigration to the United States says so. America has become a land of great wealth but obviously wealth is not the same thing as quality of life or millions of Europeans would still be immigrating to America.

How many American conservatives have ever lived in another country? If we compare America with the more socialist countries of Europe, we must remember that America has far more resources and a far lower population density than those countries yet so many European socialist nations rate ahead of America in the U.N. annual ratings of the best places to live. This does not necessarily mean, however, that America should become exactly like Europe, which to me is far too secular.

My theory of economics is simple. I call it the board theory. Suppose you have a board ten feet long and you wish to cover the maximum area with the board on a surface in two dimensions. Where would you cut the board? If you cut the ten-foot board at the seven and three point, you will be able to cover 3 x 7 or 21 square feet. If you cut the board at the six and four point, you will be able to cover 6 x 4 or 24 square feet. It is right in the middle that gives you the most coverage. If you cut at the five and five point, you can cover 5 x 5 or 25 square feet.

Economics works the same way with regards to wages and prices. Imagine one end of the board as wages and the other as prices. There is a certain midpoint at which the best balance will be achieved between wages and prices. I believe that it can best be achieved right in the middle of the political spectrum. This will produce the most good for the maximum number of people.

My theory of politics is just as simple. I call it the Bicycle Theory. When someone is riding a bicycle there are times when they would lean to the left or right, such as when they are going around a corner. But the vast majority of the time, the rider would be straight up. So it is with politics. Sometimes it is better to lean to the right, such as when a new land is opening up and people can start a settlement for themselves and there is not the established structure to provide social benefits. Sometimes it is better to lean to the left, such as when capitalism self-destructs in a stock market crash or there has been some other type of disaster. However, just like on a bicycle, the most efficient system is in the center.

The central point of my idea of socialism is that workers should be guaranteed a certain wage. It is imperative that anyone working full time be able to support an adult and one child above the poverty level. Plainly and simply, the economy booms when people have money to live on. Our problem is the vast number of business enterprises that try to get away with paying workers as little as possible.

Henry Ford realized that if workers have money to spend, they will buy things and business owners will prosper. It really does not require a rocket scientist to figure that out and he paid his assembly line workers considerably more than he really had to. Ford was a kind of anti-robber baron. He had the right idea and Republicans have tried to discredit him by pointing out his alleged anti-Semitism.

Republicans seem to think that tax cuts for the wealthy benefits society because they will reinvest the money and thus create jobs. The trouble is that most of the resulting jobs tend to be for little more than minimum wage. My theory is simple, an average person working at an average job should be able to live an average life without going much into debt, except for long-term loans for houses, cars, etc. Obviously, you can see that this is not the case today. Necessities of life are way too expensive relative to wages for the average person.

The result is that today there are tens of thousands of Americans living in homeless shelters even though they have full-time jobs. The so-called "invisible hand of the marketplace" that capitalists put so much faith in often does a poor job of coordinating wages and prices. Simply because it is much easier for a store to alter it's prices than it is for a worker to get another job while he is already working full-time.

Remember that the root cause of the 1929 stock market crash was low wages, warehouses were full of goods but the workers who made the goods did not earn enough to be able to afford to buy them. Factories thus began cutting back on production and it spiraled into a devastating crash.

Let me tell you a tale of two stores. One store paid it's workers wages that they could actually live on. But the store down the street paid it's workers as little as the owner could get away with. The result was that the workers in the first store could afford to buy the goods in the second store but not vice versa. In a very real way, we could say that the owner of the second store was actually stealing from the owner of the first store.

If every workplace had to pay workers a wage they could actually live on, the entire economy would benefit simply because people would have money to spend. If you are a business owner, you would have to pay your employees a livable wage but you would more than make it up because other workers would have money to spend on your goods.

Capitalism produces inequality, but in doing so supposedly creates a meritocracy. Except it does so very inefficiently. The object of moving to the left, to the political center is to give capitalism maximum efficiency.

Consider a sum of money, five hundred dollars for example. If you were to put this money into the social pyramid, where do you suppose it would do the most good? One of the many less-wealthy persons near the bottom of the pyramid or one of the few wealthy near the top? It is a nonsensical question, of course it would do more good if given to someone near the bottom of the pyramid.

But on the other hand if we just give out money like that, it will destroy the incentive that people need to work hard. The process of creating a civilized society is to strike an effective balance between these two factors. That is what socialism strives to do. It rewards hard work but does not allow the people with the advantages to set everything up to suit themselves.

Consider the use of water by a farmer to irrigate his field. He must get the water to the field. But to just flood the field would be destructive. That would drown some of the plants on one side of the field and parch those on the other side. Instead, the farmer uses an irrigation system to effectively distribute the water.

Consider the use of fire to provide warmth. If we are cold, all we have to do is torch the house and we will keep warm. However, it would be rather destructive. Instead, we could use a fireplace or furnace to contain a controlled fire and distribute it's heat usefully.

In a nuclear reactor, we do not allow the reaction to just run wild and melt down. Instead, it is moderated and controlled to provide continuous and sustainable energy.

Capitalism works the same way. It is a productive system, to be sure, but it can be very destructive if allowed to run wild and brings many social problems. Capitalism should be controlled like water or fire to provide the most benefit for the most people.

Capitalism operates a lot like an internal combustion engine. To keep the engine running efficiently, it requires just the right amount of air and fuel in the mixture. I believe that wages and prices in the economy works similar to the air and fuel in a car engine.

Republicans tend to think socialism is all about taxes and bureaucracy but do not usually mention the artificial needs and tremendous fluff created by raw capitalism. The "Roaring Twenties" during the era of robber barons was considered as a showcase of American capitalism until it all came crashing down. The 1980s were the Ronald Reagan days of "conspicuous consumption" until that came crashing down too. Reagan did not get the blame for the ensuing recession because it did not set in until after he had left office.

Many capitalists think that socialism is the support of people who do not want to work but do not mention the two million people being supported in America's prisons because capitalism breeds so much crime.

Some Americans would be very surprised to know all that is socialism around them. Remember that the capitalism of a hundred years ago was very raw and callous. It is socialism that has civilized society during the Twentieth Century.

Did you know that mandatory public education is socialism? If the robber barons had had their way, you can be sure that children would be laboring for pennies at an early age instead of being in school.

How about safety regulations and minimum wage laws in workplaces? In the era of robber barons, such ideas were practically blasphemy. The reason that communism went so far is the abominable conditions that so many millions lived under in western countries. I do not hesitate to say that it was socialism that addressed the grievances put forth by communists and saved the west from communism. I believe that it was socialism that really won the Cold War. The world of the robber barons would have faced massive communist-inspired revolt if not for the widespread implementation of socialism.

It is actually socialism that gives us the middle class and prosperity of today. Capitalism alone will produce a few rich and many poor due to the wealth spiral.

Wealth naturally attracts wealth and so the rich inevitably get richer while the poor get poorer.

It is bizarre how much absolute faith capitalists put in the power of private enterprise when all around them, the basic necessities are run by the government. This includes the education system, police and fire protection, as well as the military and sanitation. It was the ideas of socialism that replaced the Tammany Hall corruption of a century ago with civil service exams.

There are so many things that are beyond the bounds of the capitalists' invisible hand of the market. Three examples are the development of the atomic bomb, the space program and, the internet. All of these were simply too big for corporations to delve into without the government. Also, it requires a government rather than a private enterprise to perform tasks that are necessary or beneficial for the nation but do not offer much immediate potential for profit.

It is my opinion that these all-American ventures, the bomb, the space program and, the internet, as well as the national highway system can accurately be described as the fruits of socialism. A basic definition of socialism is government involvement in the economy. One might counter that government power gets abused, but corporate power is just as vulnerable to abuse as can be seen in the 2002 corporate scandals.

As far as the internet goes, it is fortunate that it was developed by the U.S. government instead of corporations. If the internet had been developed by rival corporations, there would have been a myriad of competing standards instead of the universal standards that the government set down.

The "invisible hand" does not handle what I call level two enterprises. Suppose that there was a beautiful flower garden along a walkway in a business district. No, the flower garden is not making any money. Yes, the garden costs money to maintain. According to capitalist logic, the garden should not exist.

Yet, suppose the garden made the area more attractive and contributed to drawing people to the area who spent money in nearby businesses. Suppose it increased the quality of life in the area. Flowers are known to have a calming effect on people and Europeans know that gardens help to lower the crime rate. The garden is an example of a level two enterprise. It is worthwhile to have but according to simplistic capitalist logic, it should not be allowed to exist because it is not making a profit.

Many people argue that private enterprise is more efficient than the government in running a business and I agree. But the government has the advantage of scale, some things are simply too large for corporations. As well as the basic fact that the business of government is the welfare of it's citizens while the business of business is to make money.

Those Republicans adamantly opposed to government involvement in the economy seem to forget the New Deal. It was government programs, in other words socialism, that had to rescue capitalism when it fell flat on it's face in the stock market crash.

One of the reasons that capitalist countries seem to have a little more economic growth is that workers in socialist countries tend to work fewer hours than those in the more capitalist nations. But, what about the ill effects that this brings? Excessive working hours takes a toll on personal relationships and families. The U.S. government has issued guidelines on nutrition and exercise recommendations, but with all these hours of work when is a person supposed to do an exercise routine or take time for proper nutrition?

It is vital for people in today's world to be informed and well read. The whole idea of the industrial revolution was for people to work less, read more and thus contribute still more new ideas to society. But this is very difficult when working overtime.

Excessive working hours is the root of a multitude of evils that those trumpeting capitalist productivity usually neglect to mention. How many times, when reading the story of the latest kid to go wrong and make national headlines, have you heard the line that his parents tried hard to provide a good home but were always working and so did not see as much of their son as they should have?

Do you recall the story of Kitty Genovese? In the spring of 1964, she was stabbed to death near her home in Queens. The thing that horrified the nation was that 38 people witnessed the random crime and no one made any attempt to help. The crime occurred as the number of hours that people were working was rising quickly. I believe that this increase in working hours detracted from the community to the point where none of her neighbors could relate to Kitty enough to want to get involved.

Taking a vacation goes far beyond just relaxing nowadays. This is an era of ever-increasing internationalism. We live in a global village. It is not good for the world for people to stay ignorant of other cultures. I believe that one should see different countries and hopefully get to know the people and possibly even learn the language. Of course a country like America is going to be disliked if other people take the trouble to learn about us but we could not be bothered to learn about them.

The trouble is that in capitalist countries like America, the vacation time for most workers is far less than in socialist countries. In most European nations, four weeks annual vacation is mandatory by law. People in socialist countries are inevitable better informed and familiar with the world outside their own countries. The primary reason is fewer working hours, thus more reading time, and more vacation time.

Remember that long working hours may show up favorably on a nation's productivity indexes but there is a heavy price to pay. Most capitalists do not realize this. The most obvious way to make it easier for people to learn new skills outside of work is to reduce their working hours. If you ask how we can keep society going with less man-hours that would be easy if we worked on reducing fluff.

In a nation like America, owning a car is practically essential. Bus service or other public transportation is extremely spotty in most areas outside of large cities. I believe that this is very wrong. Unless one chooses to live way out in the countryside, owning a car should be an option rather than a necessity.

The fourth major component of my idea of socialism for America after living wages, reduced working hours and, a mandatory minimum living standard is livable public transportation. Anyone should be able to get to and from work by public transportation without undue hardship or delay unless they live deep in the countryside.

Automobile ownership, maintenance and insurance take a big chunk out of most people's budgets. Why should a student, trying to concentrate on studies, have to work at least 20 to 30 hours a week primarily to pay for keeping a car on the road because we live in a society that has allowed car ownership to become a practical necessity? Why should a low-income family have to shoulder the burden of paying for a car because working would be very difficult without a car to get to work?

A vast amount of disposable income that could be spent on books, clothes, restaurants, vacations, home improvements, etc. instead has to be spent on paying for a car. American Republicans like to criticize the system and the taxes in the more socialist countries in Europe but usually neglect to mention that citizens of those countries are usually free to get by without a car. Do not get me wrong, I like cars, many of my ideas have to do with cars, but car ownership should be an option and not permitted to become a necessity.

The destructive effects of the automobile could fill a large book, so-called inner cities are a product of the automobile, the fewer cars that we have on the road the better. Almost certainly, the greatest single source of air pollution in the world is American automobiles. Many thousands of people die every year in car wrecks. The sleep deprivation caused by long working hours combines with the car culture equals fatal accidents. The whole world wonders about America's motives for involvement in the Middle East, is it truly for democracy or because of it's insatiable demand for foreign oil?

About a million animals, on average, die every day on America's roads. I have always hated seeing animals that have been run over. Deer are magnificent and elegant creatures and were not meant to die under the wheels of a car and lie dead by the side of the road or be splattered all over and color the road red with blood.

I recall a dog lying dead by the side of the thruway in Hamburg, NY, it's eyes were still open wide and glazed with the agony of being crushed by a vehicle. It's fangs were still bared as it had desperately tried to stop the tire that was crushing it.

Another time, in rural Niagara County, NY, a hapless kitten had wandered into the road and gotten hit. It's brains had been forced by the pressure of the vehicle's tire right out through it's eye sockets. The kitten was lying in the road with it's brains all over it's face.

What better way to teach a child how cheap life is than to let them see such road carnage? Automakers can fit a whistle device onto the front of a car to give off a supersonic sound when the vehicle is moving to warn animals but apparently they cannot be bothered.

Automobiles have brought about a sprawl spiral. We need a vast amount of parking space for cars. This causes buildings in the suburbs to be built further apart to provide room for parking. This further enhances the need for cars. This requires more parking space, which requires that buildings be built further apart, which further enhances the need for cars, which requires more parking space. It is kind of a ridiculous spiral isn't it?

The automobile-induced sprawl of cities requires a larger and thus more expensive infrastructure and services. I am not saying that we should tear everything down and build anew, only that we should begin to reverse the destructive effects of the automobile. Have you ever flown over a suburb and noticed the tremendous amount of space that must be devoted to parking and how it bloats the size of the city?

The typical American suburb is built for cars, not for people. While doing sales work, I was alarmed at how difficult it is to walk in many places. The fact that people might want to walk must have been the furthest thing from the planner's minds. In many business areas of suburbs, there is not even a sidewalk and this has led to a number of tragic deaths. Not to mention the general feeling of alienation that such an environment can produce.

Speaking of alienation, what is an older person supposed to do in a car culture when they get too old to drive? In July 2003, an 86-year old man was driving in

Santa Monica because he had no other way of getting around. He drove into a crowded outdoor market and killed ten people. At first, he did not realize what he had done. He said he may have stepped on the gas instead of the brake.

One of the focal points of my version of socialism is the bus. Buses have tremendous advantages over cars. You can read on the bus. Even if people do not talk much on the bus, you notice who gets on and off at which stop and this provides a sense of community that cars have destroyed.

If we could begin to make our way back to a society designed for walking, that would be even better. Walking is exercise and it would begin to curb the obesity epidemic. Canadians and Americans are the heaviest people in the world thanks to the automobile. People can interact or at least say hello when walking to work.

Next on the agenda is ownership of guns. This right to gun ownership in America is absolutely insane. In the U.S. in 2002, over 28,000 people died of gunshot wounds. In the town where I live, three people were just shot to death over the Fourth of July weekend. The U.S. government is always complaining about nations such as Mexico for not clamping down on drugs heading for the U.S. market. But what about those nations in which gun ownership is illegal, many of those nations' citizens manage to get their hands on American guns. The gun used in the assassination attempt on Pope John Paul was an American-made Browning handgun. The U.S. government does not seem too concerned.

As so often happens with capitalism, a destructive spiral has gotten started. Some people have guns and there is inevitably some gun violence. This leads more people to buy guns for protection, which leaves more guns lying around and thus leads to more shootings. This causes still more people to buy guns for protection, which leaves still more guns laying around, which leads to still more shootings. The only ones that have ever benefited from this is the industries that manufacture guns and those who hate America and look for evidence of what an evil society it is.

On several occasions, foreigners in American cities have been confronted by police looking for someone else, the foreigners, being unsure of their language skills reached for their wallets to show identification. Unfortunately, the cops were used to the ways of a gun-loving society and assumed that the foreigner was reaching for a gun. The cops proceeded to fill the foreigners with bullets.

The standard NRA line is that "Guns do not kill people, people kill people". Whatever, if the gun was not there it would not have happened. It is people with guns that kill people. It is many times harder to kill someone with a knife than it is with a gun.

Meanwhile, what is the world supposed to think of the United States, which is trying to show the world how wonderful democracy is? For God sakes, the high school in my town has to have a police station in the school.

It is long past time to put an end to it. No more guns.

As far as justice, there is always more crime in a capitalist society because of the harshness and gross income disparities. America has, at the time of this writing, two million people in prison and that is not even counting juveniles and those on parole and probation. In an almost comical situation, George W. Bush was recently lecturing Iran on human rights when the fact is that an American has four times the chance of an Iranian of being in prison.

I believe that the death penalty must be abolished in America. In 1976, the death penalty was brought back after being banned several years earlier. Republicans were delighted, promising that this would drastically reduce serious crime. Did it reduce serious crime? The fact is that the number of murders in the U.S. has multiplied about three or four times since the death penalty was reestablished.

Look at the other advanced countries, none of which has the death penalty, how does their murder rates compare with that of the U.S? If the death penalty advocates were right, those countries should have murder rates far above that of the U.S., instead the opposite is true. The death penalty is just sending the message that the way to solve a problem is to kill people.

The death penalty must be abolished if for no other reason than the many mistakes that are made. On many occasions, an investigation into a convict's case has proven, often after the convict has spent many years in prison, that he could not possibly have committed the crime. Often, DNA evidence that was not available when the trial took place was the deciding factor.

In 2003, the governor of Illinois commuted all death sentences in his state because of the many mistakes. A while after that it was announced that the six men who had spent fourteen years in prison for the notorious 1989 attack on the Central Park jogger were being released because the real attacker had confessed. A while after that, a federal judge resigned because the justice system is just wasting lives.

In New York State, we have what is known as the "Rockefeller Drug Laws". Former governor Nelson Rockefeller introduced the laws in the early 1970s. The laws proscribe extremely harsh penalties for small-time drug offenses. This does not mean that someone caught with marijuana should not be punished, but destroying a life with twenty years in jail is not the answer. The laws have only destroyed the lives of those at a relatively low-level on the drug pyramid while leaving the big-time dealers virtually untouched. And, of course, given the condition of the judicial system we have to wonder how many innocents are in jail because of this.

Most people agree that the United States justice system is badly in need of reform. Those defendants fortunate to be able to afford a good lawyer have a far less chance of being convicted than those forced to rely on public defenders. There have been numerous stories in the news of public defenders falling asleep during a serious trial because they have been up all night working on other cases or being so overloaded with cases that they could not remember the name of the defendant that they were arriving to defend.

Clearly, the spirit of capitalism has found it's way into the courtroom and justice is for sale to those who can afford a good lawyer. I shudder at how many innocent people have already been executed. This is utterly against everything that the United States of America is supposed to stand for. I believe that we need socialism to put things right.

In 2002, U.S. News and World Report reported that U.S. military court martials, for a person in the military accused of a crime, have an incredible 96% conviction rate. For enlisted persons, it is closer to 98%. How on this earth can a legitimate court have a 96% conviction rate? This is unbelievable.

Animal rights also have a long way to go. The idea of throwing a live lobster into a pot of boiling water or a steaming oven is barbaric. Even before it's death, lobsters commonly spend days packed together with other lobsters. For such a solitary and territorial creature, this in itself must be torture. What if lobsters were cute and furry, would this be allowed to happen?

Does anyone know how much a calf has to suffer to produce veal? Why are animals like elephants, bears, lions and, tigers allowed to spend their lives doing ridiculous tricks to entertain people? Circuses want people to think that the animals are having fun but you can be assured that they are treated very harshly to get them to perform the tricks. Many elephants have made desperate attempts to escape such a life. Why is this allowed? This is terrible.

In 1994 Tyke, a 21-year old African elephant, made a desperate attempt at freedom during a performance with the Shrine Circus in Honolulu. She killed her trainer, who had been fired from a previous job for ill-treating animals and was found to have alcohol and cocaine in his system.

Cops fired bullet after bullet at Tyke, who wanted nothing more than her freedom. She ran through the downtown with blood on her face from the cops' bullets. Finally, Tyke collapsed in the street from the effects of the cops' bullets in her head. She died a slow and miserable death, which paralleled her life, as the cops kept firing bullets into her. She thrashed about with her trunk for quite some time. The body of a magnificent creature, meant to be roaming the wilds in freedom, was dumped in a landfill. I recall watching this spectacle on the news and I do not ever want to see such a sight again.

Tyke had made several desperate attempts previously to escape the horrible life she was forced into but still she was kept in service. Why is this allowed?

Capitalists and conservatives actually remind me of the lobster industry. The industry is making money and naturally wants to keep it that way. They keep telling everyone that the lobster does not really feel much pain, if any, when it is thrown into boiling water. I do not believe it. Conservatives say that yes, capitalism creates hardship but everyone is better off as a result. I do not believe that either.

Lets look back to the beginnings of civilization. This is when people banded together to escape the law of the jungle. Capitalism has the same spirit as the law of the jungle; the rich get richer while the poor get poorer. Socialism has the spirit of civilization; give people incentive but guarantee them the bare necessities of life. Wealth naturally flows toward wealth. Socialism moderates it to benefit everyone but never far enough to damage incentive.

Capitalism is not the best system. It results in large numbers of poor and unused talents because people cannot break out of their situations. It results in the destructive cycles that we have looked at-wealth concentration, education advantage, guns and, sprawl. It inspires productivity and innovation but it's overall design is very wasteful. It produces a cold and nasty culture. The incessant bombardment of advertising leads everyone to imagine that "the whole world revolves around me".

Political conservatives point to the taxes and "bloated bureaucracy" of socialism but neglect to mention the fluff, crime, prison population and social problems that come with capitalism. They also do not mention the vast amount of social benefits, such as compulsory education and workplace safety codes that are from socialism.

They talk of this "invisible hand of the marketplace" but neglect to mention how poorly it handles wage and price coordination (because it is naturally much easier for a store to alter it's prices than it is for a worker to get another job). Or how poorly it handles low-cost housing (in how many American cities is there a glut of office space but a shortage of affordable housing?). Or how the invisible hand does not handle large-scale research or projects at all.

What about drug prices? Americans pay the highest prices in the world for prescription drugs. I recall when I was doing a sales campaign in an Erie, PA K-mart. I chatted with a man coming into the store that had just gotten out of hospital following his recovery from a heart attack and was coming to the store's pharmacy to fill his prescription. He seemed to feel very victorious. But I saw him also as he left, he could not afford the medicine that he was supposed to get.

What about unemployment benefits? At the present time in New York State when someone loses a job and applies for unemployment immediately, it is usually about eight weeks before he gets his first unemployment check. Suppose that a man is supporting a family but living paycheck to paycheck. He suddenly loses his job through no fault of his own. What is he supposed to do for eight weeks? If anyone would give this some thought, they may realize that this is probably where a lot of crime gets started.

During my years as a Republican sympathizer, I noticed more and more how the thinking is very short-term and there are usually hidden long-term costs to Republican concepts. Take, for example, electricity deregulation. This was a conservative idea to bring down prices by competition and it seemed to be working. I used to sell alternative electric and gas supplies to homeowners.

What we were not being told was that the utilities did not have enough money coming in with the lowered prices to modernize the archaic and overloaded electrical grid. Now, after the Blackout of 2003, billions of dollars are going to have to come from somewhere to fix the system. Long-term thinking is usually not a Republican forte. This is the same kind of short-term thinking that produces so much fluff.

Republicans rarely mention that the United Nations almost always chooses socialist countries for it's top positions when publishing it's annual rating of the best countries in the world to live. According to the U.N., the top twenty in quality of life in 2003 were (in order): Norway, Iceland, Sweden, Australia, Netherlands, Belgium, USA, Canada, Japan, Switzerland, Denmark, Ireland, Britain, Finland, Luxembourg, Austria, France, Germany, Spain and, New Zealand. As you can see, with the exception of the USA and possibly Japan, the list is extremely socialist.

How can small, densely populated countries with few resources provide health care and a social net for their people but America cannot. Please do not say that socialism does not work because it does. It is the vested interests of capitalists that makes them not want it to work.

Things were going fairly well in America during the 1990s. As soon as George W. Bush came into power, the economy headed down. Socialist Italy usually placed between ten and fifteen in the U.N. rankings. In 2001, Italians uncharacteristically voted to the right. Now, after two years of Silvio Berlusconi's conservative policies, Italy is nowhere to be found in the top twenty. Canada was in poor shape under conservative Brian Mulroney, the turning point was the election of liberal Jean Chretein in 1993 and Canada proceeded to lead the U.N. rankings for eight years in a row.

Socialist countries produce better students, take better care of the environment and, have better read and informed people. America is known for it's poor student performance, particularly in geography, math and, science. Persons with leftward political beliefs tend to be smarter. In America, look at the intellects of presidents Kennedy, Carter and, Clinton compared with Reagan and George Bush Jr.

Reagan is known for expressing the beliefs that ketchup is a vegetable and that trees cause air pollution. Just before becoming president, George W. Bush was asked if he could name the leaders of six certain large foreign nations and he could name only one, that of Mexico. George Bush Sr.'s vice president, Dan Quayle, is remembered for his unsuccessful attempt to spell "potato" while visiting a grade school class during a spelling bee. Albert Einstein was interested in politics and was a devout socialist.

In these days of terrorism, socialist governments are better equipped to handle emergencies. A socialist government could keep Amtrak operating, even if it was not earning a profit, just in case it was needed as an alternate means of transportation to the airline industry. If we just operate on capitalist principles, industries that cannot earn a profit will fold, regardless of the potential value of such industries to national security or emergencies.

For one thing, I believe that America and Britain should keep a certain amount of steel industrial capacity intact regardless of market forces. A nation is not really independent without access to steel production capacity.

The problem with socialism is not that it does not work. The rankings of the U.N. and the lack of European immigration to the U.S. says it does. The problem is that socialists have had a history of making too many promises. In no way do I promise that my version of socialism will create a paradise, it will merely create a better society than we have now.

We will not have a paradise until Jesus returns to establish his kingdom on earth and then all this politics will be irrelevant anyway. I want to emphasize right now that following the Bible will bring more improvement than any political ideas.

My basic formula for life is: Be liberal in politics, be conservative in religion.

Corrupt and unrealistic unions have given socialism a bad name. I am the first to admit that some of the socialist literature that I have read is completely nonsensical and unrealistic. While it was titled "socialist", it was in fact, far left, or communist. The way to preserve jobs for people is not for unions to make unrealistic demands that are incongruent with today's world but to reduce working hours to give people a chance to upgrade their skills.

Socialism has also been the party of sinners. In American politics, Nixon represents the Republican capitalist abuse of power while Kennedy and Clinton, although they had leftist politics that worked, represented the personally sinful lefty. Socialists have often been atheists.

Who says that the right rather than the left is the party of God? The fact is that the Bible does not endorse any political system. It could be that capitalists wanted simple people to believe that the existent order is God-ordained. Just as capitalism had time to recover and get it's act together following the devastating 1929 crash, and it did so by embracing many socialist ideals, the new socialism will be the moral party of God.

I want nothing more than for our society to get rid of it's suffocating secularism and rampant immorality. I want this much more than I want socialism itself. What I really want to do is put together Christianity and socialism.

Keep in mind, that while the socialism of the 1970s can be considered to have "failed", it never failed anything like capitalism did in the 1929 crash. Remember also, that when I describe socialism as "left" I mean left relative to where America is now (2003). I believe that the best in politics and economics is to be found in the center.

As far as I am concerned, both capitalism and communism have collapsed. Capitalism collapsed in 1929 and communism in 1991. The truth is that far-left communism actually lasted longer than far-right capitalism. It is the center that works and America still has the ghost of the robber baron era and is significantly too far to the right. The object is to change this.

A person in a free society has three things, wealth, rights and, opportunity. What we want is not equality of wealth. That would destroy incentive. We want equality of rights and equality of opportunity. If we have this then society will be at it's best politically. This is what creates the greatest productivity and the best quality of life.

After the stock market crash in 1929, capitalism was given the chance to fix it's faults and make a comeback. It did so by incorporating socialist ideas. It is now about time for the new socialism to make it's appearance.

Have you noticed that every thirty years or so, socialism reappears in America? In the thirties, it was the New Deal. In the sixties, it was the Great Society. We may be seeing the next incarnation of socialism in America with the disgust over the corporate scandals of 2002 and the widespread anti-globalization protests. I would like to name the next manifestation of socialism in America "The Just Society". A primary goal will be to drain the cold, callous nastiness brought about by capitalism.

Capitalism produces wealth, we all know that, it just brings too many evils along with it. The modification of capitalism, the movement to the political center, is what we need. If we cut the board right in the middle, we will have a good life for all. If you are a business owner, you should be looking forward to this as much as anyone for it will ensure that people will have money to spend on your goods.

The new socialism must guarantee a minimal living standard for all, just enough to exist on. A person cannot realistically hunt for a job in today's society if he cannot afford a phone, transportation and, a place to live. Somewhat reduced working hours will enable people to study in the evenings in order to advance.

The new socialism is not about supporting people who do not want to work, it is about people who work full time while living in homeless shelters. It is not about coddling criminals, it is about ensuring that those convicted are in fact criminals by ensuring that justice is equally applied. It is about working fewer hours and taking longer vacations not because people are lazy, but so they can have time to be informed, improve their skills, start enterprises of their own and understand other countries. It is not about government interfering in business, it is about covering those vital areas that the capitalists' "invisible hand" does not and reducing the out of control fluff that provides jobs but no lasting benefits to society.

Universal health care is mandatory. Many Americans go bankrupt with credit card debt when a medical emergency hits and there is no health insurance. Many people have to work into their late seventies to survive and have adequate health coverage. It makes America seem barbaric and draconian in comparison with the other advanced nations.

Capitalism concentrates on production but handles distribution very poorly. Communism concentrated on distribution but did not handle production well. Socialism is in the middle and handles both very well.

Conservatives understand that people need incentive to work hard. What they do not understand is that those with the advantages have a tendency to try to set things up to suit themselves. Communists were the opposite, they knew from watching the robber baron societies of the west that those with the advantages like to set everything up to keep it that way. What they did not understand is that people need incentive to work hard. What we need is the center that understands both.

A socialist party does not even have to win to be effective. The presence of a socialist party forces the other parties to address the issues of quality of life. The other parties know that if they do not address the issues, voters will flock to the socialist party. It is time for a society that is kind, compassionate and, just instead of cold, callous and, nasty.

IDEA # 234; FLOURESCENT CHALK: I'll bet that this would be a novelty. Chalk that is fluorescent and thus glows in the dark. It may contain particles that absorb light and re-radiate it or may simply contain small reflective particles that shine when exposed to light, depending on the purpose the chalk is to be used for. This would be far from just a toy and would have many industrial and site applications.

IDEA # 235; RENAISSANCE MAN CONTEST: We have long had mental contests like spelling bees and trivia shows as well as physical contests such as track and bodybuilding. However, this is the Twenty-First Century now. A man today is expected to be good at many things. So much knowledge is available and so many skills necessary that it is time to encourage generalization. Too much specialization is too limiting. Thinking that a man should be good at only one thing is very commoner.

I propose a Renaissance-Man contest. There will be a series of events, some intellectual and some physical. Every year will be crowned a Mr. Renaissance. Events might look like the following:

1. Sketch a drawing of a given subject.

2. Trivia resembling "Jeopardy" science, history, literature, technology, world events.

3. Write a story in at least one foreign language.

4. Mathematical calculations and problem solving.

5. Take apart a motor and reassemble it.

6. Various feats of strength.

7. Physique.

8.   Endurance and stamina exercises.

9.   Running race.

IDEA # 236; ESTIMATING DISTANCES IN SCHOOL: One useful skill that schoolchildren get almost no training in is the estimating of distances. All that is necessary is to paint lines around the school and schoolyard. Every child should have a good idea of horizontal and vertical distances and angular measures. A ten-meter line could be painted down a hallway. A vertical five-meter line could be painted on the outside of the school. A square meter could be illustrated on a floor. A twenty-degree angle could be painted in a circle on a wall.

IDEA # 237; COLLEGE COURSE ON CORPORATIONS: We live in an age of globalization in which traditional nation states are becoming less significant while multi-national corporations are becoming more significant. Students learn all about the world's nations in school. When are we going to get around to teaching a class about the world's big corporations?

IDEA # 238; TWO HANDS OF TEN KEYBOARD: Standard computer keyboards are beginning to look a bit archaic to me. This would be a replacement for the traditional qwerty keyboard. It would consist of two pods of ten keys each. Each hand would be placed on a pod. By pressing one key on each pod at the same time, one hundred possible keys would be available. The two pods could be placed in any position so as to improve comfort. I believe that this would virtually eliminate carpal tunnel syndrome.

IDEA # 239; ARROW CHARACTER ON COMPUTER KEYBOARD: Why is it that computer instructions often contain arrows? Arrows are also frequently used in scientific writings. Yet, there is no way to make an arrow with a conventional keyboard.

IDEA # 240; AUTOMATIC MESSAGES BY STRING OF CHARACTERS ON KEYBOARD: In some businesses such as banks and those in which there is a significant amount of cash, there is sometimes a "hidden button". This is for employees to press if a robbery or some other emergency should occur. Today, in the computer age, why not modify this? Let's set up the computers in an office so that if a designated character is held down for several seconds, it will set off the alarm just as the old hidden button would.

IDEA # 241; UNDERWATER SWIMMING POOL MIRROR: This would be a novelty. Placing a fairly large mirror on the side of a swimming pool under the water level.

IDEA # 242; FAN MUFFLER: Many cars have dual exhaust. The reason for this is that it makes it possible for the engine to work better by eliminating the waste gases produced by combustion faster. Since the faster the car exhaust is eliminated, the better, why not look for further ways to accomplish this?

At present, cars use only the action of the pistons to push the exhaust gases out. How about placing a small fan at the end of the exhaust pipe. The fan could possibly be electrically powered and the blade could turn at high speed. This would pull the exhaust gases out of the engine even faster, thus increasing efficiency at minimal cost.

IDEA # 243: TUNE INTO SOUNDS AT A CERTAIN DISTANCE BY WAVE CURVATURE: Every sound wave coming your way has a certain curvature. When a sound comes from further away, the wave will be flatter than a sound coming from close by. This is because the distant sound wave will be part of a larger circle. This can easily be observed by throwing a stone into a pond. The further from the source the wave moves, the larger will be the circle and thus the flatter it's edge.

We can use this concept to tune into sounds coming from a certain distance and tune out those coming from closer or further away. Set three microphones equally spaced on a horizontal boom perpendicular to the source of sound we are seeking. The middle microphone will be on a smaller horizontal boom that is perpendicular to the first boom and thus pointing straight at the source of sound. The microphones will be set electronically so that to be accepted into the system, it must enter all three microphones at exactly the same instant. This can be done because any sound has a definite frequency and amplitude, distinct from other sounds.

Next, set the middle microphone so that it is movable going back and forth on it's boom. The two end microphones will be fixed in position and not movable on their booms. By moving the middle microphone back and forth when the small boom that it is on is pointing directly at the source of sound, we can make it so that only a given wave curvature will hit all three microphones at the same instant. This sound thus will be the one received into the system. When we move the middle microphone forward, it is only slightly back from the two fixed microphones and will pick up a flat wave from a distant source. When we move the microphone back, it will pick up a much closer sound with a more curved wave, at the same instant as the two fixed microphones.

IDEA # 244; RETRACTABLE AIRCRAFT WING: The wings of an aircraft provide lift to raise the plane off the ground. The wing is shaped so that the air must travel faster over the top of the wing than the bottom. This creates more pressure below the wing than above it.

An aircraft going at a certain speed has an optimum wingspan. Too little and not enough lift is provided. Too wide and too much drag is created when the airplane is moving fast. Altitude is also a factor. Air is much denser at lower altitudes and so a shorter wingspan will provide more lift than it would at higher altitudes. There is also more drag in the denser air at lower altitudes, so too wide of a wingspan would not be desirable.

What all this means is that an aircraft could be made considerably more efficient if it had a variable wingspan rather than a fixed one. The wingspan could be shortened at low altitudes and widened at higher altitudes. It could be widened at low speeds and shortened at high speeds. At any given altitude and speed, the most efficient wingspan could be provided.

Until relatively recently, this probably would have caused more difficulty than it would have been worth. However with the extremely strong and light materials available today, I believe that it is time for the retractable aircraft wing. The most obvious solution would seem to be a telescopic wing section at the end that could be extended and retracted as desired by hydraulic power.

IDEA # 245; THE FORGOTTEN FLAVOR: When is someone going to come out with a watermelon juice or soft drink? I think that it will be very good. As far as drinks go, watermelon seems to be the forgotten flavor.

IDEA # 246; GLASS WITH THE SAME REFRACTIVE INDEX AS WATER: I believe that this is a forgotten aspect of safety glasses. What if glass, such as that in a car windshield, had the same refractive index as water? The refractive index of a transparent substance is the ratio of the speed of light in that medium to the speed of light in a vacuum. Water on the outside of the glass would be, optically speaking, a continuation of the glass. The objective is for water on the window to not interfere with vision.

IDEA # 247; FLUORESCENT COLOR PAINT ON CURBS: Why not paint curbs on the side of the street with fluorescent colors? This would not only look futuristic but would be a safety factor. Each street could have it's own color.

IDEA # 248; FOREARM WARMER PAD: Suppose you work in an office in which the temperature is a little cooler than you would like. It is not possible to bring a heater or to turn up the heat. The solution is forearm warmer pads. Place the pads on your desk and put your forearms on the pads. The heat will conduct through your body. It will not be as ideal as having the heat on but it will be considerably better than nothing. The pads can be powered by conventional electric power but could also be chemically powered like the hand warmers taken on hunting and camping trips.

IDEA # 249; LIGHT COLOR FOR ELECTRIC WIRES: Electrical resistance in a wire is dependent on a number of factors, including the temperature of the wire. Heat increases electrical resistance in a conductor so that a colder wire has lower resistance to the electric flow.

Across the world are many billions of kilometers of electrical transmission wire on telephone poles and the large metal utility towers. The wire that carries the current seems to be invariably black in color. Black absorbs heat. This means that the wires are warmer than they would be if they were a lighter color, especially in the daytime under the sun. Hence, there is more electrical resistance and so more power loss in transmission due to the resistance of the wires.

Power companies do all they can to minimize power losses during the long transmission journeys such as stepping up the voltage at the expense of current by a transformer. Why not start making the wire in a lighter color? It would not be much of a savings over a short distance but the saving would mount up over time into a fortune.

IDEA # 250; ACOUSTIC BURGLAR GLASS: Burglars sometimes break glass to gain entry. What no one seems to have ever thought of is the creation of a glass that makes a loud a noise as possible if broken for use in windows.

IDEA # 251; PAINTING COMPLEX OBJECT BY WATER LEVEL: Suppose we have a situation in industry in which complex objects of some kind are to be painted. Is it not possible to mix a paint that would float on water and then paint the object by simply lowering and raising the water level? Multiple colors could be painted on either by masking or by painting in one color and then moving the object in such a way that would then make painting with the other colors possible.

IDEA # 252; QUICKLY ABSORBED FOODS FOR AMUSEMENT PARKS: Amusement parks with wild rides should serve foods that are absorbed quickly by the human digestive system. This would leave the riders with less in their stomachs to vomit.

IDEA # 253; WATER EXERCISE: I believe that the water in a swimming pool offers an excellent resistance for exercise purposes. Imagine two flat pieces that look like wings and fit onto a person's arms. All kinds of motions are possible to thoroughly exercise all muscles of the upper body. The water wings could conceivably have holes that can be opened or partially closed or entirely closed in order to vary resistance. In fact, this exercise device will probably make it possible to work the muscles from different angles than conventional exercise equipment.

IDEA # 254; GRAVEL UNDERGROUND WALL TO PREVENT PRISON ESCAPES: In a number of old World War Two movies, prisoners of war escape by digging a tunnel under the fence. Why not dig a trench around the areas where the prisoners are housed and fill it with gravel? This would make tunneling out impossible.

IDEA # 255; TAPE OF BARKING DOGS: Different breeds of dogs have different barks. I am sure that dog lovers who buy books about dogs would also like a tape of dog breed identification by bark.

IDEA # 256; GLASS DIGESTIVE SYSTEM EXHIBIT: We should be able to make a glass or clear plastic model of the human digestive system. If we could add mechanical teeth and all the appropriate chemicals, it would be possible to make a working model of the digestive system. This would add a lot of realism to studying a medical textbook.

IDEA # 257; AMBIDEXTROUS TRAINING KIT: When someone injures their favored hand so that it requires a period of recovery, it is usually necessary to learn to make more use of the non-favored hand. For most persons, this will be the left hand. Why is there not a kit with simple exercises and intricate movements designed to improve skill in the non-favored hand? It could be based on the experience of thousands of patients in learning to make better use of the formerly non-favored hand following injury.

IDEA # 258; ABSTRACT MOVIE: You know what abstract art is. It is a modern art form that focuses on the feeling rather than literal realism. If we can move from the literal to the abstract in painting, why not do the same in movies? The technology available today makes possible an entirely new form of movie.

The movies that we have had so far are all literal. What we can do now is move on to evoking feelings and expressing thoughts and ideas through creative use of colors and shapes. The movie may have some literal or semi-literal images but will be mainly a "color bath" for the visual sense. A skilled artist will be able to convey a journey through colors, patterns and shapes. I believe that this will take much more imagination than literal movies.

7

GROOVED-IN THINKING

Just what is it that is stopping us from thinking of new ideas? It is as if we are walking a path through a field. We tend to stay on the beaten path rather than try to go where no one has walked before. We tend to automatically do things the way they have always been done. Even when we do look for new ways, sometimes our assumptions at the beginning of an endeavor were not the best. Imagine that there were new people from a distant land who knew nothing of our traditions and way of life but did learn our technology. They would certainly see things that we do not see.

Also notice how many people who make discoveries are self-taught. This avoids the grooved-in thinking certain to come with a formal education. An ideal example of breaking out of grooved-in thinking is the digital clock. Who said that a clock has to be round?

IDEA # 259; FACING DOWN BED: We usually think of sleeping lying on the back or on the side. This is what beds have been designed for. It is time for a bed setup for sleeping facing downward. It will require a gap in the mattress so that the sleeper can face downward while lying on the stomach. I am convinced that this may help to remedy certain sleep problems. Not to mention cases in which injuries or sunburn make sleeping on the back difficult.

IDEA # 260; MODERN CLOCKS: The clock is open to big improvements in design. Digital clocks are just the beginning. The ancient design of a clock is modeled on the turning earth. Why not have a clock using light to tell the time instead of the old mechanical hands. Then it would act as a nightlight too. A blue light beam could be for hours and a red one for minutes, sweeping over the face of the clock.

Who says a clock must be circular? What about a bar clock with a scale of sixty at the top of the bar and twenty-four at the bottom. A red point of light goes from one end of the bar to the other in an hour to indicate minutes and a blue point of light does the same thing in a day to indicate hours.

IDEA # 261; MILITARY RANK BY SPECTRUM: Soldiers have used the chevron patches to indicate NCO military rank for a long, long time. Maybe it's time for a more modern look. Why not use the colors of the spectrum to indicate rank? From red to blue, vivid patches of different colors and shades could be used to show rank instead of the archaic-looking chevrons.

IDEA # 262; NEW ABBREVIATIONS: I got to thinking about our language. Why is it that we can abbreviate first by writing it 1st. We can abbreviate one by using it's number symbol, 1. But there is no way to quickly write single, double, triple, quadruple and, so on.

Since those terms all end with 'le', why not use that for the abbreviation? Henceforth, single can be written as 1le. Double can be written as 2le. Triple can be written as 3le. Quadruple can be written as 4le. It should eventually be possible to have the 'le' as small and elevated letters in 1st.

IDEA # 263; HELICOPTER TOW TRUCKS: This would be a breakthrough. Sooner or later it will be cost effective on packed highways. It would greatly shorten traffic backup when an accident happens on a crowded highway in a big city. In such cases, the tow-truck must somehow find it's way through stalled traffic and thousands of cars can be idled for over an hour.

The large helicopter could easily and quickly pick up the car by electromagnet and possibly deposit it on a nearby road where it would not be blocking traffic and could easily be retrieved by a conventional tow-truck. The electromagnet could be coated with plastic so that it would not damage the car.

IDEA # 264; TRAFFIC LIGHT ELIMINATION: Most city roads were built when traffic volume was much lower. When road upgrading is done, why not look to building an overpass and underpass at really busy intersections. This would cost money but would eliminate many hours of wasted waiting, wasted fuel and, pollution by idled vehicles. I believe that as much as possible should be done to make this an urban trend. With the overpass and underpass, cars could just zoom through the intersection without a traffic light unless the car was turning. This would really be a great thing.

IDEA # 265; CONVEYING FEELING IN PRINTING BY LIGHT OR DARK TYPE: With word processing programs it is quick and easy to change type; whether size, boldness or font. I recall high school print shop and in the old days this would involve painstaking setting of type. Naturally, we developed a tradition of just using standard type for printing and just keeping things as simple as possible.

We are no longer bound to limitations like this. Why not give each character in a story his or her particular font? Whatever a character says will be written in their particular font. Just as people's voices do not sound the same, their fonts will be different.

It is very easy to make larger or smaller letters, bolder or fainter. A whisper could be written in smaller letters, a shout in larger ones. Or, each subplot could be written in a different font. Maybe large or bold letters could be used for events going on close by and small or faint ones for events happening at a distance.

A little creativity and boldness will make it so that writing a story involves not only writing but will be like painting a picture as well. The extra meaning that can be put into writing with the tools available in standard word processing programs is unlimited. All that is limiting is the grooved-in way of writing developed over hundreds of years of archaic printing presses and techniques. All we need to do is break out of these old ways.

IDEA # 266; TEACHING LANGUAGES BY GROUPS: If you climb up a tree to get a piece of fruit out of a bunch, you may as well take the whole bunch after putting in all the effort of climbing up. So it is with most languages.

We live in a time when there is more people than ever who speak more than one language but also more need than ever for more language skills. Most languages are part of a group. Germanic languages include English, German, Dutch and, the Scandinavian languages, except Finnish. Romance languages include French, Spanish, Italian, Portuguese and, Romanian. Slavic Languages include Russian, Polish, Serbo-Croatian and many others.

If someone learns a language, they already know the basic patterns of the other languages in the group. Many languages of the same group are so similar that the speakers can often make out what the other is saying even without any training in the other language. With this in mind, it seems almost wasteful to learn a language without taking the time to begin learning the other languages of the group.

IDEA # 267; ADJUSTABLE ACCELERATOR PEDAL: I have noticed that it is easy to acquire a 'lead foot' when changing cars. I had one car in which it was necessary to really push down on the gas to go. Then, I got another and I only had to touch the accelerator for the same effect. The result was speeding tickets. Why should not an accelerator be adjustable? I think that it would really cut down on speeding.

IDEA # 268; RECTANGULAR LENSES FOR CAMERA: Look around a camera store and you will see something that is very incongruous. Except that you will probably not notice it because of thought conditioning. It just hit me one day and the fact that we do not notice it is one of the most classic examples of grooved-in thinking.

Why is it that photographs are usually rectangular but camera lenses are almost all round? This is grooved-in thinking going back centuries. The logical shape for a photograph is certainly rectangular. The logical shape for lenses is circular, going back to early lenses for eyeglasses and telescopes going back to the Sixteenth Century. In the late Nineteenth Century, cameras came along.

By this time, three hundred years of lens making said that lenses should be round, after all the eye is round. Apparently, no one considered that the camera would be most efficient if the lens was the same shape as the photograph was to be. There are rectangular lenses for magnifying glasses that are the same shape as a photograph. However, no one seems to have applied this logic to cameras.

Let's have a camera with the lens in the same shape as the photograph. This would be much more efficient. Using a circular lens to gather light to produce a rectangular photograph sounds like the old cliché of a mismatch, "fitting a square peg into a round hole".

IDEA # 269; THE POWER OF TWELVE: I have long had an interest in the Metric System. It seemed so logical, the decimal convertibility of units was much easier on students and saved them the trouble of memorizing all of those odd numbers to convert units in the old system. Most of the world was metric and I thought that surely the U.S. would go officially metric soon.

Recently, it just hit me all of a sudden what a failure the Metric System really is.

I now believe that there is something alien about the Metric System. Something is just missing about it. I have come to the conclusion that the Metric System just does not fit neatly with the way people naturally do things. If you look at the history of the Metric System, you will see that despite it's apparent advantages and widespread support and promotion by governments, it still usually had to be forced on people by law.

Another thing that puzzled me is how the Old English System just refuses to die. (By the way, I am not exactly sure why the use of traditional units like miles, feet, inches, etc. is called "The Old English System" since England is metric, but since that is what it is commonly known as, that is what I will call it.)

I did some thinking and noticed that the English units are designed for fractions, while the Metric System is designed for decimal multiples. The twelve inches of a foot is easily divisible and the thirty-six inches in a yard, sixty seconds in a minute twenty-four hours in a day and, three hundred sixty degrees in a circle are nice, round easily factorable numbers.

The English system also had what I will call "natural units". The length of a meter (also spelled metre in many countries) is basically arbitrary. It was originally intended to be one ten millionth of the distance from the equator to the North Pole on the Paris meridian. The English units such as inches, feet, yards and, miles were lengths that one regularly used in daily life. A needle or a button was about an inch in size. A knife or hand tool was about a foot long. A shovel, a saw or, a space sufficient for a person to walk through measured about a yard. The diameter of an average town was around a mile.

A nautical mile was a great example of a natural unit. It is somewhat longer than a statue (or land) mile. A nautical mile was based on the circumference of the earth and was one minute of one degree at the equator, in other words, one sixtieth of one three hundred sixtieth of the earth's equatorial diameter. This comes out to 1.16 statue mile and was very useful because sightings on celestial bodies were used for navigation.

By far, the quantity that we measure the most is time. Human beings measure time probably far more than all other quantities combined. Every time you look at a clock or watch, you are taking a measurement of time. The main reason that I have decided the Metric System to be a failure is that since just after the Metric System was introduced, there has been no serious effort to measure time in metric.

Have you ever wondered, if the concept of the Metric System is so great, then why do we not measure time in metric?

For that matter, angular degrees never went metric either. Thus, latitude and longitude is not expressed in metric. There is a decimal system of measuring degrees in so-called grads instead of degrees but it never got far.

The incredible truth is that even outside of the U.S. in devoutly metric nations, only a small percentage of all measurements are done in metric units.

The main reason is that time measurement and angular measurement has not been converted to metric. There are still the old sixty-second minutes, sixty-minute hours, twenty-four hour days and, twelve-month years. Decimal seems nowhere to be found.

This illuminates the fact that we have two distinct "number cultures". There are four basic arithmetical operations that people do on a regular basis in daily life: addition, subtraction, multiplication and, division. Addition and subtraction is easy in any measurement system. The so-called English System is very useful with division due to it's use of nice, round, easily divisible numbers. The Metric System is the opposite. It offers easy unit convertibility based on multiples of ten but does not fit readily with commonly used fractions such as thirds, fourths and, sixths.

The two distinct number cultures that we deal with are "the decimal" and "the factorial". We could also call the two; "multiplication" and "division". If we wish to divide a meter into thirds, we are circumventing the entire purpose of the system if we express it as 1/3 meter. The proper metric decimal alternative is a messy .333333...meters. The strength of the Metric System is the number ten, but it's weak point is the number three.

The problem with metric is that outside of temperature measurement and the scientific and engineering communities; fractions are, if anything, more important than multiplication with regards to measurement units. Measurements of time and latitude and longitude never went metric at all because natural units and easy division is far more important in these fields than the decimal conversion factors that the Metric System offers.

Put simply, in metric it is easy to multiply a unit because units are based on multiples of ten. But if one wishes to divide while using metric units, it is a bit more problematic. It is easy to divide a metric unit by either two or five. But what if one wishes to divide metric units by three or four or six? The system designed to make things easier suddenly becomes more difficult. Metric was not designed with division or fractions in mind and in the measurement of time or latitude and longitude, we rarely require a multiple of a unit but often require a fraction.

The importance of fractions to human beings can be illustrated by the fact that even in U.S. currency, which operates on a decimal model, the quarter has never gone away. If decimal was really superior we would see no need to keep a quarter dollar, 25 cents, in circulation. We could have a 20 cents or 30 cents piece. I find the quarter as evidence of how vital fractions have always been to human thinking and dealings. The Metric System tries to decimalize everything and does not take fractions into consideration. That is it's great weakness.

Metric is at it's strongest in temperature measurements. The temperature scale is nothing but a straight line where multiples and fractions are practically irrelevant. Metric also works well in the scientific and engineering communities. Scientists and engineers often carry calculators around with them, or can do complex mathematics in their heads, so the loss of easy divisibility with the metric system is unimportant. This is a large part of the appeal of the Metric System, it sounds so "modern" and "scientific" and this is what I thought for many years until giving it further thought.

Metric works better in so-called "straight line" measurements. It is a little weaker in dealing with weight and length than it is with temperature because these measurements may require division by common fractions that metric is not designed for. But still, metric can be probably be considered as stronger than the English System in measurements of weight and length because of it's decimal convertibility, even though the English System has more natural units.

The zone where metric completely washes out is the so-called circular measurements. That of angular degrees, latitude and longitude and most importantly, time. If everything was a straight-line, metric would reign supreme, but it isn't. Considering that the most important and most frequent measurement humans take is time, the Metric System must be considered as far less than successful. The decimalization of time is almost inconceivable.

The Metric System was designed to make measurements easier by revolving around ten and water (for weight measurements). It has ignored fractions and factors other than ten. Since common fractions have been essential to humans since prehistory, the Metric System cannot really be considered as achieving it's objectives. This is regardless of how it is spread around the world.

It's proponents have ignored the fact that time, by far our most frequent and important measurement, has not and will not be decimalized. The dominant measurement system is the one that handles the measurement of time and we are measuring it in the same way as in centuries past.

This failure of my beloved Metric System caused me to give more thought to the way in which the world is divided between the two number cultures, that of decimal and that of factorial. Even if the Metric System is a "failure", the old system of feet, inches and, miles, with their awkward conversion factors, is not actually ideal for the modern world either.

In fact, the Metric System was created over two hundred years ago in an attempt to address the flaws in this old system. It is a testimony to the weakness of the Metric System that this old system is still hanging around in the Twenty-First Century.

This really got me thinking some more. In our high-tech world of today, why should such a simple issue as measurement units come with all of these complications? Why is there not a measurement system that can handle everything efficiently? One would think that issues such as these would have been solved by the end of the Eighteenth Century.

I continued thinking about the concept of a measurement system that can handle all measurements efficiently and easily. My thinking led me to a stunning conclusion. The reason that attempts to develop an efficient and all-around measurement system inevitably splits into the two number cultures is our base ten number system. We have basically been counting all wrong for thousands of years.

The Metric System and it's inter-convertible units is a brilliant idea. It's great flaw is that it cannot handle the vital common fractions without becoming more awkward than the system it was designed to replace. I decided that the flaw is not with the system itself but with our use of ten as a number base.

This can only mean that the base ten system, used by the world for thousands of years, was far from the best choice.

Human beings have dealt with numbers since pre-historic times. Counting probably began with dots or marks. Later, figures and names for numbers were developed; the familiar Arabic numerals 1,2,3...

The number system can go as high as infinity. The only trouble is that we have to find a name and symbol for each number. It is possible to take a shortcut and reuse the symbols. To do this, numbers go up to a certain level and then reuse the symbols by remembering to add the amount of the level that we have taken the symbols up to.

A convenient base would be a compromise between not having too many symbols and not being too unwieldy. A base thirty system would require too many symbols and a base four system would be too unwieldy.

Since human beings have ten fingers and often used the fingers for counting, this seemed a logical number of symbols on which to base the number system. So today, we have ten characters in the number system; 0,1,2,3,4,5,6,7,8,9. When we count higher than ten, we reuse the characters again so that we can have numbers like 14, 59 and, 36,481,342. This is known as a base ten system. Each column counts as ten.

A number system does not have to be based on ten. The computer industry uses a hexadecimal or base sixteen number system because it is convenient to have each symbol in this system describable by four bits of data. The computer itself only understands 0 and 1 so it works on a base two, or binary, system. A base four system, for example, would use only four characters or; 0,1,2,3.

A number system of any base can be used for arithmetic. We just use a base ten system because it was convenient for ancient men to be able to count on their fingers. We also use numbers for listing things such as addresses, not just for figuring. However, if we put listings aside, we see that some numbers find their way into use much more than others.

The truth is that this base ten system that has been with us since ancient times is a hindrance that we do not often stop to think about. It is so deeply rooted in human life that we do not notice how inefficient it was since the beginning. No matter how brilliant an idea the Metric System is, it is still based on ten. If ten was the most efficient base for our number system, the Metric System would be much more successful than it is. It is vital to integrate our counting and measurement systems. Apparently, no one has yet noticed this.

We may have ten fingers but decimal does not fit with those nice, round numbers that humans have always found so useful. No scientific analysis was ever done to find which base would make the most effective number system. We are in a base ten system for no better reason than ancient finger counting. A base ten system is efficient for counting on the fingers but not much else. The only reason that multiples of ten crop up as much as they do is because the decimal system makes it artificially convenient to do so.

In the opinion of this writer a base ten system is a poor choice except for counting on the fingers. The plain truth is that numbers naturally revolve around twelve, rather than ten. Failure to change the number base is what has so undermined the Metric System.

In fact, without thinking about it much, we have been torn between ten and twelve for centuries, just as we are torn between the two number cultures in measurement. Notice that when we come to the teens, we do not begin calling numbers -teen until thirteen, the number after twelve. The suffix -teen means "ten", so why do we wait until after twelve to begin using it? Why do we not call eleven 'oneteen' and twelve 'twoteen'? I believe it is because we know, at least instinctively, that twelve is the natural focal point of the number system.

The element carbon seems to be the focal point of the periodic table, forming many more compounds than all other elements combined. In fact, life is based on carbon. The number twelve is kind of like the carbon of numbers. Just by coincidence, the atomic mass of carbon is twelve.

Consider two women named Molly and Sally. Molly is in charge at work because her family owns the company. She is a decent and capable person. Sally, however, is the kind of person that everything seems to naturally revolve around. Every community has a Sally. She is not necessarily assertive or seeking popularity, she just seems to be made to be the center of things and that is how it is.

In the community of numbers, ten is like Molly. It seems logical to use ten as a base for the number system because we have ten fingers, just as it seems logical to have Molly in charge because her family owns the company.

But twelve is like Sally. No matter what base we use for the number system, when we start to actually use numbers we find it to be a kind of revolving point. Even though we have ten fingers, our numbers naturally revolve around twelve.

In my opinion, twelve should have been the base of our number system. Everything from measuring to calculating would be easier.

We cannot get away from those nice round numbers that are so easy to use. The zodiac has twelve constellations. Jesus had twelve apostles. The Jews had twelve tribes. Heaven has twelve gates and twelve foundations. One of our favorite measurements is the dozen. The year has twelve months.

Multiples of twelve are all around us too. Jewelers use twenty-four karats. Days have twenty-four hours. Minutes are made up of sixty seconds and hours are made up of sixty minutes. A circle has three hundred sixty degrees. The opposite direction is one hundred eighty degrees. A mile is 5,280 feet. The walls of heaven measure 144 cubits in height.

Actually, there was once ten months but two more were added to make twelve, July and August. We just cannot get away from this number. It will appear everywhere that human beings quantify things because it is such a natural focal point.

A base ten system does not coordinate the widely used common factors with the decimal system. This is what causes the great fault in the Metric System. The factors and multiples of twelve, however, are right in the middle of the factor pattern. There are two vital factor lines for daily life; the twos and the threes. Twelve catches both of them. Sixteen would also be a possible base as it is very divisible by two and it is used in computer hexadecimal but it misses the three-factor line.

Ten is only factorable by two and five and the multiples of ten are not that useful except for forty and sixty. The factors of twelve are two, three, four and, six. Eight and nine relate to twelve better than to ten because 8 x 3 = 12 x 2 (24). Eight does not relate to ten until we get to forty. 9 x 4 = 12 x 3 (36). Nine and ten do not relate until we get to 90.

The multiples of twelve are extremely useful. Twenty-four is a widely used number, such as hours in a day and jeweler's karats, because it is a very factorable number, by two, three, four, six, eight and, twelve. Thirty-six, forty-eight, sixty and, seventy-two are all widely used numbers because of factorability.

The reason that this factorability is so important is that stacking, organizing and, arranging things are so important in human life and represent so much of what human beings use numbers for. Ten and it's multiples are generally poorly factorable, two and five being the only exceptions. Eggs are sold by a dozen, instead of by ten, because a dozen is more likely to be evenly divisible among a group of people.

Fractions are very important to humans but most commonly used fractions do not fit readily with the decimal or percent system. One-third, one-sixth, one-ninth or, one-twelfth does not fit readily into decimal or percent. All of this will be changed with a base twelve system.

We have been suffering for centuries from what I have called the "factor gap". We gravitate toward nice round easily divisible numbers because of their useful-ness. But our heritage is the base ten system and that does not coordinate with these favored numbers easily.

If we had a base twelve system, the factor gap would be closed and the favored numbers would be multiples of the base number like 20, 30, 40 are today. This would make calculations and expression of numbers easier, student would have an easier time and, there would be fewer mistakes made. Common fractions would fit neatly with "decimal".

The Metric System was a brilliant idea, but it turned out to be only a Band-aid solution because it's creators did not introduce a new number base. We can never create a measurement system that can accommodate both easy multiplicity and easy factorability as long as we have a base ten number system.

The base twelve system will be very beneficial to education. Just like words revolve around vowels, numbers revolve around their own "vowels", numbers such as twelve. I define a "numerical vowel" as a nice, round, easily divisible num-ber such as twelve and it's multiples.

I have wondered, numbers are a representation of reality. So, why can a land-scape be considered as beautiful but a numberscape is usually considered as cold and devoid of charm? I believe that it is because a painting or a good photograph revolves around a focal point but the numbers do not.

Because our number system is not base twelve, it resembles a poorly focused photograph. It is "out of focus" because it is not based on naturally favored, easily factorable numbers. Maybe this is why few children are really interested in math.

There is no doubt that a base-twelve system will make expression of numerical values easier and reduce mistakes. A base-ten system can be compared with a manual transmission vehicle in the same way that a base-twelve system would compare to a vehicle with automatic transmission. Our favored numbers and the multiples of the base number would be one and the same.

Stacking and organizing things are what numbers have always been used for on a daily basis. A product sold in a twelve pack would be much more likely to be evenly divisible among a group of people. A product grouped by twelve is much easier to deal with than one in groups of ten. It can be stacked or distributed 2 x 6, 3 x 4 or, 1 x 12.

In nature, so much is shaped like a circle or sphere. In this, measurement in degrees, we have always used round numbers because of the many possibilities of even divisibility. Rectangles and cubes, which are so vital to the things that humans make from boxes to street plans, have either four or six sides. Both are multiples of twelve but not of ten.

Of course, we are blinded to all this because we have been made to think in terms of ten since childhood. It wrongly seems as if the basic counting numbers cannot get any more efficient. The folly of the Metric System was it's construction on a foundation of ten.

Metrication and computer binary and hexadecimal has introduced the mindset for conversion to base twelve. There would be a period of conversion with numerical values listed in both systems. We will need two more characters for digits so that the base twelve system will use 0,1,2,3,4,5,6,7,8,9,x,y. Except that I do not favor using x and y for numbers because of possible confusion.

In base twelve, 12 will be written 10. 24 will be 20. 36 will appear as 30. 48 will be written as 40. 144 will be 100. Things will be a lot easier. People will be linked to their naturally favored numbers much better. In base twelve, a number such as 13 (equivalent to 15 in base ten) will not be pronounced "thirteen", it will be pronounced as one-three base twelve, at least during the conversion period from base ten to base twelve.

When we get more into exploring space, these favored numbers will only increase in importance. Light-years, the distance light travels in a year, are based on time. In measuring time, we have ignored ten completely and have twenty-four hour days and twelve-month years.

Distance in space is also measured in parsecs. A parsec is 3.26 light years. Parsec stands for parallax-second. From a distance of one parsec, the distance from the earth to the sun, 93,000,000 miles, will occupy an angular distance of one parallax second of arc. You know that a complete circle is divided into 360 degrees. Each degree is divided into sixty minutes, which is divided into sixty seconds, the same as with time measurements.

The nice round numbers, the multiples of twelve, will certainly dominate space.

One tremendous possibility with a base twelve system is the expression of time in decimal. Actually, it will not be decimal because the prefix deci-means ten or a base ten number system so I'll call a base twelve system "bidecimal". We cannot imagine how easy it will be to express something as important as time in bidecimal. We have those nice, round units for days, hours, minutes and, seconds that do not fit with our unfortunate base ten system.

In bidecimal, an hour will be .05 day. A minute will be .02 hour. A second will be .02 minute. A month will be .1 year. It would be easy to do away with writing hours, minutes and, seconds and just express everything in bidecimal days. Now we are starting to sound really space age.

Places on the earth's surface will also be easily expressible in bidecimal. Each and every spot on the surface of the earth will be expressible with a number such as 156.294862. The number of significant figures will be determined by the accuracy desired. The position of a streetlight on the surface of the earth will obviously require several more significant figures of accuracy than the position of a city. It is really amazing that we have the GPS system to pinpoint any spot on earth but we are still unable to express a point on earth in decimal. During space travel, the same system can be used to describe any location on a planet.

At this point, we do not even have an efficient universal measurement system. We take only a portion of our measurements with the Metric System. For the most important measurements such as time, we are still measuring as we were centuries ago. The Metric System is actually so inefficient that the centuries old system of feet, inches and, miles is still hanging around. Yet, this old system is just not suited for modern science and engineering.

With a number system based on twelve, this division or "factor gap" in our number system would come together. All of the quantities that we commonly measure, including time and angular degrees, could be expressed conveniently in bidecimal. Most common factors would fit neatly into this system, unlike in the base ten system. It is difficult at this point to imagine how much easier this will make things. All counting and calculating will be easier and less prone to errors. More children will take a natural interest in math.

No matter how much ingenuity we may put into devising a system of measurement, with ten as the number base it is not possible to close the factor gap, to reconcile the two different number cultures that we have now. Every time you look at a clock or watch, you see numbers that go up to twelve. Time is by far our most important measurement. If ten were the most efficient number base, you would see numbers that go only to ten on a clock.

The number twelve rules, there is even a religious significance to this. So many things in the Bible come in twelves. Jesus, the Son of God, was sent to reconcile us to God. To me this is symbolized by the reconciling of our two "number cultures". Followed by the creation of a much better system by adapting twelve as the base.

IDEA # 270; RE-THINKING OUR SYMBOLS: In ancient times, it was well known the great difference that a choice of symbols can make. Over time, a symbol can work it's way deep into the consciousness of a tribe or group of people. This has absolutely nothing to do with magic or the occult. In modern times, it seems as if the power of symbols has been forgotten. Many nations today have adopted animals or elements of nature as their national symbols. Let's take a long-term look at a few of these choices.

America has chosen the eagle as it's symbol. As one might expect, it was America that invented and developed the airplane, our national symbol compels us to find a way to fly. Eagles usually build their nests in high places such as in mountains and sure enough, it is America that became known for it's skyscrapers made possible by the invention of the elevator.

An eagle is also a ruthless predator that looks for prey, sees what it wants and takes it with no regard for any creature but itself. And as one might expect, America has a crime problem many times that of comparable modern nations. Millions of Americans are now in jail or on parole or probation or dead for acting like the national symbol. I suppose that I am the only one that is incredulous that there is the image of an eagle in America's courtrooms overlooking judges sending people to prison for behaving like the national symbol.

The symbol of Britain is the lion. The British navy dominated the world's seas for four hundred years just like the lions dominate the grasslands of Africa. Lions usually only hunt when hungry but will not allow other cats like leopards and cheetahs to hunt in their territory. This resembles imperial Britain adding territory to it's empire before France or Spain could take the territory for their empires.

Lions are community-oriented creatures with a rigid hierarchy that work together for a successful hunt. Life in Britain is a model of this that would make any lion feel at home. The downside of such a community is that there is a rigid hierarchy that makes sure everyone knows their place and puts limits on individual freedoms. Traditionally, this has led millions of Britons to seek out places where there is more freedom but where they can still remain British. Much of the talents and energies that could have been applied to Britain have instead gone to build nations such as Canada, America, Australia, New Zealand and, South Africa.

Also, lions like to roam wide-open spaces. Large tracts of countryside have been preserved in crowded Britain. But just like so many lions, the lure of the Canadian prairies, the Australian outback and, the South African veld has led many Englishmen to take their energies and ambitions away from their homeland.

Canada has chosen a peaceful, benign symbol for itself, the maple leaf. And sure enough, Canada is a fine model of civilization. However, the maple leaf is not without it's disadvantages. There is nowhere in Canada to go to escape from snow. Just as the leaves of the maple tree fall off in winter, the Canadian snowbirds take hundreds of millions of Canadian dollars to vacation spots in the sun. These travelers and their money leave Canada every winter just as surely as the leaves fall from a maple tree. Having a maple leaf in the sub-conscious and the culture almost demands that one leave the tree when winter arrives.

Symbols are so powerful that even the images on a national flag can have a profound effect. America was the first nation to reach another celestial body and has been the most influential nation in space exploration and to a lesser extent, modern astronomy. Maybe we can trace this to lead to the prominent field of stars on the U.S. flag, it almost pulls Americans toward space.

Another prominent accomplishment of America is the coast-to-coast transportation networks, first railroads and then superhighways. Take a look at the stripes on the U.S. flag. Don't these stripes seem to be demanding a coast-to-coast network by spanning the flag from one side to the other? Unfortunately America has a very high incarceration rate, maybe this is related somehow to the resemblance between the stripes on the flag and prison bars.

What about the flag of Britain, the Union Jack? The blue field seems to represent the oceans while the corners and midpoints of each side are connected to the center of the flag by stripes. This very closely resembles London at the center connected to outposts of empire at the far corners of the globe. This flag is practically a plan for a worldwide empire.

The Soviet Union had the hammer and sickle as it's symbol, standing for traditional industry and agriculture. The country held up fine as long as the world was in the days of traditional smokestack industry. As soon as that period passed into history, the country followed. I wonder what would have happened if the Soviet Union had had the computer mouse as it's symbol instead.

What kind of world have we had over the past hundred years? Unfortunately, it was very much a time of war. Maybe part of the reason for this is the symbols that the nations have chosen. The symbol of America is the bald eagle. The symbol of Germany (aside from the swastika) is a different kind of eagle. The symbol of Russia is the bear. The symbol of Britain is the lion. The symbol of China is the dragon. This sounds to me like some kind of zoo from hell and we really should not be too surprised at the level of warfare in the past century.

Why do nations not choose the dog as a national symbol? Dogs are known as man's best friend and have been faithful companions of humans for thousands of years. Why is it that humans choose vicious wild predators who could not care less about humans for their symbols but this wonderful creature has been ignored?

Do eagles guard our homes and bark at intruders to alert us? Would a lion put it's own life at risk to protect a child? Would a wolf make a career out of rounding up sheep for farmers? Would a tiger spend it's career guarding human beings? Cats have been companions too but can you imagine a finicky and self-centered housecat leading around a blind person?

Maybe there is no better example of what is wrong with the world. An intelligent, capable and, loyal animal emerges from the wild and becomes the companion of human beings. You can be sure that the affection of a dog is a lot more genuine than that of your fellow humans. Yet when it comes to humans choosing animals for symbols, they ignore this magnificent creature and choose vicious wild predators that could not care less about anything to do with human beings and that would tear people to pieces in seconds if they had the chance.

IDEA # 271; ELECTRIC FLAG: A flag is made to be seen snapping in the breeze. However, the majority of the time there is little or no breeze in most locations. The solution is a flagpole with an artificial breeze. Holes or a slit can be made in the flagpole where the flag is mounted so that the airflow from a fan or set of fans at the base of the flagpole exits the hollow flagpole via the holes or slit and thus provides a breeze for the flag. The fan could be turned on or off by an electric switch.

Imagine the aura of superiority it would impart if in front of the United Nations building, the U.S. flag was seen on television snapping proudly in the breeze while the flags of the other nations were lying dormant in apparent obeisance.

IDEA # 272; PHONE BOOKS: If anyone who prints phone books would like to make them easier to use, why not make print each letter on a different color of paper. I have noticed that some people have difficulty with things listed alphabetically.

Another example of how our thinking takes years to catch up is the issue of cell phones and phone books. We are moving toward cell phones and away from landlines yet few cell numbers are listed in phone books. The problem is that cell phones are often transitory. A lot of people tend to have their phones shut off or to go with another company and switch to another number.

The solution that I favor is the use of the internet for online phone books. There could be online "open" phone books in which anyone could enter in their phone number in those towns from which he was likely to be called. The names would be automatically alphabetized and entries could be deleted or numbers changed at any time. When someone first enters a listing, he would choose a password so that no one else could change his number or delete his listing.

IDEA # 273; ADJUSTABLE PHOTO BACKGROUND: People come in different colors, the hair as well as the skin. In those places where people have photos taken for passports, drivers licenses and so on, why not make the photo background screen in a choice of colors? The color screen could be selected that would contrast the most with the subject's skin and hair color pattern.

IDEA # 274; RETRACTABLE LIGHTS: Do you want to know what would make a room look really ultramodern? Lights on the wall or ceiling that retract during the day and come out when turned on.

8

DE-INVENTING

I am so interested in getting more out of existing technology that I want to introduce the concept of de-inventing. This is the art of finding a low-tech solution to a problem that is just as effective but much less expensive than a high-tech solution. It is the use of simple common sense rather than high-tech wizardry. I believe that we are just too impressed with high technology.

The perfect example of de-inventing is a dog. We can create high-tech burglar alarms but an old-fashioned dog will do just as well, if not better.

Here is a few examples.

IDEA # 275; ELEVATOR SHAFT AS BUILDING AIR PUMP: I wonder why elevators are not used to circulate air in large buildings. The elevator going up and down the shaft is a piston moving vast quantities of air. The elevator shaft is usually closed off but I believe it could do a great deal to circulate air through the building. The elevator will apply positive pressure when approaching and negative pressure when moving away. The resulting pressures could be used to circulate air either through the hallways of the building or air ducts.

IDEA # 276; BALL AND RING AIR MOTION DETECTOR: This is a low-tech burglar and fire alarm. It will be far less expensive than other such alarms. It centers around a large but very light object such as an inflated ball. The ball is suspended from a support of some kind by a metal wire. There is a small metal ring around the wire at some point but not touching it. The metal ring and the metal wire are part of a circuit. If the metal wire touches the ring, it closes the circuit and sets off an alarm of some kind.

When the alarm is turned on and a movement of air causes the ball to move and thus, the wire to touch the ring, the alarm goes off. The ring could possibly be moved up or down the wire to vary the sensitivity. When a burglar or fire causes a movement of air, the alarm goes off.

IDEA # 277; FLAT AIR MOTION SECURITY DETECTOR: This is similar to the Ball and Ring Motion Detector except that it uses a picture in an open frame that is open and the picture attached only at the top. The picture will be held taut in the frame by a metal strip attached to it's bottom. However, the metal strip will be hanging loosely at the bottom of the picture and will not be in contact with the frame.

When a movement of air causes the picture to move, the metal strip moves too and comes in contact with metal in the frame. Both are part of a circuit. The circuit has then been closed and an alarm is set off.

IDEA # 278; CLEAR, HEAVY PLASTIC SHEETS TO MEASURE AREA: Suppose you have a map and a certain very irregular-shaped area of the map that you would like to know the area of. Or what if you have a graph and wish to find the area under a curve without complex calculus?

The solution is clear, heavy plastic sheets of uniform thickness. Simply place the sheet over the map or graph and mark the outline of the area that you wish to measure. Then cut out that section carefully with a small, sharp knife. Next, weigh the cut out section on a sensitive balance or scale. Now that you know how much it weighs, all you need to do is find out how much a piece of the plastic weighs that covers a known area.

Several trials can be done to increase accuracy. The larger the piece of plastic used in the cut out, the less the chance of error due to minute variations in the thickness of the plastic sheets.

IDEA # 279; MEASUREMENT OF ELEVATION IN HILLY COUNTRY BY SMALL AIRCRAFT SHADOW: To map elevations in hilly country, fly a small plane over the country on a sunny day around noon. Keep the plane at constant speed and altitude, relative to sea level. Take a continuous video of the plane shadow as the journey progresses.

From this video it can be determined the distance of the shadow below the plane by the size of the shadow. Since it will be known how high above sea level the plane was, it can easily be calculated how high the terrain is at that particular point. Since we also know that the plane was kept at constant speed, we can calculate the exact point that fits the particular elevation.

9

POSSIBLE NOW

Technology is advancing so fast that we barely begin to get the most out of it. There are a myriad of uses that we could be getting out of recent technology but have not noticed yet. Our minds have not caught up yet.

This applies especially to materials science and electronics. Think of ideas that people had in the past but were not possible because materials were not available that were strong enough or light enough. We may have these materials today and forgotten ideas could be revived.

IDEA # 280; THE COMEBACK OF CALISTHENICS: Not that calisthenics ever really went away. Supposedly, the scientific weight training of today has replaced the old calisthenics that were a staple of physical culture in days past. With the advanced exercise machines in health clubs today; the old push-ups, chin-ups, jumping jacks, and so on look very Nineteenth Century.

But not so fast, I am all for weight training. However, there have been two important trends in recent decades. First of all, people have less and less free time. Secondly, people have gotten significantly heavier.

A workout with calisthenics requires considerably less time than one with weights. This is the major drawback of weightlifting. Less time is required for calisthenics simply because the trainee does not have to keep changing the weights on a barbell.

A calisthenics workout requires little or no equipment and thus saves the additional time that would be spent on the commute to the gym or health club. Additionally, it is much easier to get in a calisthenics workout than a weightlifting workout when traveling.

Face it, the number one reason for dropping out of a weightlifting program is time, or rather the lack of time. It is one thing for a young athlete to adhere to a weightlifting program at the gym. But what happens several years later when he (or she) has a home and family? The usual result is the end of the weightlifting program.

You may be thinking that weightlifting builds more strength than those old calisthenics exercises. You are absolutely right. However, things have changed. People are now considerably heavier than they used to be. Exercises like push-ups and chin-ups make use of the body's own weight in building strength and endurance. A fellow that weighs 140 pounds can keep his body fit but will not build a tremendous amount of strength with calisthenics.

However, what about someone that weighs 220 pounds or more? That is quite a bit of weight to be lifting with each repetition of each exercise. Gravity does not know or care whether your body is fat or muscle, in shape or out of shape. The muscles just know that they are lifting a significant amount of weight and they think: "hey, I had better add some more strength". A heavy person gets more benefit out of calisthenics than a lighter person.

I started getting serious about exercise in the summer of 1976, just after tenth grade. For about two years, I made steady progress with calisthenics. Endless push-ups, sit-ups, knee bends, and so on. After high school, I got into weightlifting and devotedly continued for a number of years. As time went on, other interests crowded in; religion, reading, writing, building stuff out of wood. I became ever more of a news addict. I became determined to really understand the world outside of my own countries.

Yet, exercise was a priority that I was not willing to give up on. I had struggled too hard to get into shape during my early days of exercise. I found myself moving back toward calisthenics. I found that I could get a really good workout in a relatively short time. I did not work out for as long as when I was more into weightlifting but I made up for it by doing a calisthenics workout just about every day instead of the usual three days a week of weight lifting.

At one point, I went a year without doing any weightlifting and I wondered if I had lost any strength. I went back to the weights and saw that I had nothing to worry about. I weighed over 240 pounds and my bodyweight was providing me with a very adequate strength workout. I had lost no strength at all since I had last hit the weights.

Alternatively, a trainee can continue with weightlifting but can make use of calisthenics for a so-called "short workout". This is another system that I used. The limiting factor to a trainee is usually time. So, do a "long workout" with the weights if you can. If not, at least do a short workout with calisthenics. That will stop the body from deteriorating until the long workouts can be resumed. Remember, never do nothing. Also, the change of workout will hit the muscles from a different angle. As any weightlifter knows, this is vital to continuous progress.

By the way, I do not believe in so-called isometric or isotonic exercises. The muscles must be taken through their full range of motion during exercise. I used to have a device called a "Bullworker" that I used for many years. Finally, I abandoned it because I realized the importance of taking the muscles through their full range of motion, which the Bullworker did not do.

An exercise program with calisthenics could include push-ups (press-ups in Britain), chin-ups (also called pull-ups) from both directions, knee bends (try doing deep knee bends super-slow), situps, twists, dips (I put one hand on a board across a sink and the other hand on a nearby dryer to do dips), push-ups can also be done with one hand just make sure to go all the way down. Work should also be done for the extremities that this program does not cover such as grip exercises or fingertip pushups for the hands and lower arms and calf extensions on a step for the calf muscles.

IDEA # 281; VACUUM FLIGHT: Around Renaissance times, the idea arose to make a boat fly by attaching metal spheres to it and pumping the air out of the spheres. Air does have weight. An automobile travelling at 30 miles per hour plows aside 60 pounds of air a second. Pressurizing a jumbo jet may add a ton to it's weight. Actually, a standard 55-gallon drum used in industry would be just about light enough to fly if all the air was pumped out of it.

Unfortunately, the idea was spoiled by the fact that pumping the air out of metal spheres would cause the atmospheric pressure of 14 pounds per square inch at sea level to crush the spheres.

But that was then and this is now. Modern materials science has created materials that are phenomenally strong. Maybe the concept of vacuum flight can be brought back into consideration. The air pressure from outside would actually help to hold the components of a vacuum craft together.

IDEA # 282; THE BUSINESS NUMBER: Identity theft is a major problem nowadays. Everyone seems to be protective of the social security number. The number was originally intended for use only by the government but corporations found it useful also. The solution is to transform the number into a business number that can be used for all the business purposes that it is now. A new numbering system can possibly be created for government use. The business number system would be designed to be safe. Your number would be useless to anyone getting hold of it.

IDEA # 283; THREE-DIMENSIONAL GAMES: It is really time for age-old two-dimensional games such as checkers and chess to join the new millennium. Games like these can be played in a clear three-dimensional box using lights instead of pieces.

The game box would be divided into the usual number of squares but in three dimensions instead of two. Each cube would have two lights in it of two different colors. A light would be on according to whose "piece" was in that cube. Aside from that, the game would be played pretty much as always.

Games such as these have been used for centuries to develop a sense of strategy, particularly for future military officers. The problem is that hundreds of years ago, warfare was strictly two-dimensional. The surface of the land or sea was the battlefield. For the past century warfare, and life in general, has been three-dimensional.

The airplane, skyscraper and, to a lesser extent, spacecraft and submarines have opened up the third dimension. Yet, the games that do so much to prepare students for the patterns in the playing field of life are still limited to two dimensions as if it were 1880. We will probably never be as sub-consciously comfortable with the third dimension as with the other two but these games should be preparing students. Gravity once confined human endeavors to two dimensions but that has long since changed.

In order to help prepare students for this reality, I maintain that games such as those listed above should be as three-dimensional as possible. This is not possible with sports. But in sports, the motion of a ball through the air provides experience with the third dimension. It is games such as checkers, chess and, board games that should strive for three-dimensionality. In fact, should have been striving for it a hundred years ago.

IDEA # 284; SOLAR SYSTEM ORBIT CRAFT MEASURING STAR PARALLAX: Astronomers have a method for measuring the distances to stars. It is a trigonometric technique known as parallax. Sightings of the star are taken six months apart. The astronomers look for a small shift in the star's position relative to the background stars further away. Since the earth is known to be 93 million miles from the sun. That would mean that the two measurements had been taken 186 million miles apart. Since this is the baseline and we know what the angular shift was, the distance to the star can be calculated trigonometrically.

The trouble with this method is that it is only accurate for nearby stars, to a distance of fifty light-years or so. The distances to stars further away must be measured by less direct methods.

The solution is a spacecraft designed to take such measurements. The spacecraft will be remote controlled from earth by radio. We will put the spacecraft in the same orbit as Jupiter but on the opposite side of the sun from Jupiter. That way, it's existence will not be threatened by Jupiter's powerful gravity. Since Jupiter is 483 million miles from the sun, that will give us a baseline of almost a billion miles to measure distances to stars.

The downside is that the spacecraft will only orbit the sun every 11 years if it is in Jupiter's orbit. So, measurements will be taken 11 years apart. However if we could have two of the spacecraft, each could be put in the orbit. One would be a quarter of the orbit ahead of Jupiter and the other a quarter of the orbit behind Jupiter. Then a measurement to a star could be taken from both spacecraft at once and it's distance immediately determined.

IDEA # 285; HIGH EXPLOSIVE LONG-RANGE GUNS: As stated previously, I really do not wish to come up with new and more deadly weapons. However, the only thing worse than coming up with more deadly weapons is to have an enemy come up with new and more deadly weapons first.

Guns have always been powered by what is known as 'low explosives'. The explosives used for demolition and in warheads are much more powerful and are known as 'high explosives'. Dynamite and TNT are high explosives. Gunpowder and cordite are low explosives.

An artillery piece uses a low explosive to propel the shell from the gun and a high explosive warhead that explodes when the shell reaches it's destination. High explosives would, of course, provide much more launching power for the shell than low explosives. The reason that high explosives are not used to propel the shell is simply that they would blow the gun apart when it is fired.

Today however, there are entirely new materials available with extreme strength that make steel seem flimsy by comparison. I believe that based on these new materials, a new and long-range gun is possible using high explosives rather than the traditional low explosives to launch the warhead.

IDEA # 286; FUTURE MILITARY UNIFORMS: There is one thing that military planners seem to be forgetting when it comes to future uniforms and gear for combat soldiers. That is color.

When it comes to uniform color, planners still seem to be in camouflage mode. That is, making uniforms green to aid in concealment in vegetation or tan and brown to aid in concealment in the desert. The fact is that conventional camouflage is much less important today than formerly due to infrared vision devices able to sense body warmth and which thus are immune to camouflage.

There is a very real possibility that tactical nuclear weapons will be used on future battlefields. What we know is that such a weapon would produce a vast amount of radiation. Dark colors absorb radiation much more than light colors. In the atomic bombing of Hiroshima, there were incidents in which a person wearing a dark kimono burst into flames while someone nearby in a white kimono did not.

In my opinion, this would mean that all future military uniforms should be very light in color. Some kind of camouflage pattern would still be possible but the main consideration would be the resistance to radiation in the event of the explosion of a tactical nuclear weapon nearby.

IDEA # 288; INVISIBILITY: Now that we have so-called stealth technology, what about outright invisibility. I believe that we now have the technology to make an object, say a dirigible or aircraft, completely invisible.

First, let's define what invisibility is. An object is invisible when a viewer will see the background on the other side of the object when looking at the object as if the object were not there. In other words, when an object is made invisible and an observer looks at the object and there is a brick wall in the background, the observer will see the brick wall as if the invisible object were not even there.

Glass is invisible because the electromagnetic waves coming from the other side of the glass will pass right through it to an observer. An object made of glass is, of course, visible because it refracts light.

Today we have digital cameras as well as flat screen television displays. By combining these two technologies, we can achieve invisibility. Suppose we constructed an object to be invisible by covering it's surface with tiny electric cells like pixels. Each cell would be both a receiver of electromagnetic waves, as in a digital camera and a transmitter of waves, as in a flat screen television. Each pixel cell would act as a mirror as it would receive light of a certain color and also transmit light of a certain color.

Except that what we would do is "mate" the cells. Each electric cell would be mated with the cell diametrically opposite on the other side of the object. There would be a data bus in the invisible object similar to that in a busy computer network and each cell would transmit that color received by it's mate on the opposite side of the invisible object. That way, light would appear to pass right through the object and by fine tuning, the object could be made literally invisible. If a similar way could be found to receive and transmit waves of non-visible wavelengths, the same could be done with respect to radar and other wavelengths.

IDEA # 289; IDENTIFICATION OF ATOMS: Do you know what would really be a major step forward, to be able to tell one atom from another? I do not mean separating an atom of hydrogen from one of oxygen because we can do that already. I mean looking at an atom of oxygen and taking note of distinguishing characteristics of that atom and then being able to find that same atom when it is with other atoms of oxygen.

There must be some kind of distinguishing characteristics of atoms in regard to the same kind of atom. Atoms of a kind cannot be completely identical. Maybe the nuclear force holding the nucleus together is slightly different from atom to atom. Possibly we can find slight variations in the electron orbitals. Is everyone sure that the energy of electrons orbiting all protons in the same levels is the same?

A new frontier is the search for a way to tell apparently identical atoms apart. Think of all this could do for us.

IDEA # 290; INFOMAN: Kids have been given so many television heroes starting with Batman and Superman. The most prominent super abilities of these heroes seem to be either super strength or the ability to fly. What about one that will provide motivation to study?

Welcome to the adventures of Infoman. The power of Infoman is in his knowledge and wisdom. He cannot fly and is probably not as strong as Superman. Nevertheless, Infoman has powers much more important and gets even better results than the other heroes. Also, his abilities are within reach to the average student. I remember how I was impressed in my youth by the professor on Gilligan's Island.

IDEA # 291; MICROSCOPIC RECORD REPAIR: Now that the old plastic records are not made any more, suppose one has a cherished old record that is damaged or scratched? What we need is a "record doctor" that will put the disc under a microscope and repair it with finely applied plastic cement.

IDEA # 292; ELECTRIC PENCIL TO SIMPLY DRAW CIRCUITS: A basic electric circuit consists of a power source, such as a battery, a load, such as a light, a wire from the negative side of the battery to the one side of the light. The circuit is completed by another wire from the other side of the light back to the positive side of the battery.

Of course, circuits can get a lot more complex than that. I got to thinking, for research and educational purposes why can we not have a special pencil that can simply draw the wires in electric circuits. Kind of like a schematic drawing that actually conducts electricity. The drawing will probably have to be done on a specially prepared surface, such as plastic, that is non-conducting. The pencil itself will have to use some material that comes off like pencil lead, actually graphite, and conducts electricity well.

IDEA # 293; USE OF EARTH'S MAGNETIC FIELD AS PARTICLE ACCELERATOR: The earth has a strong magnetic field. If you have ever seen the northern lights, that is caused by the fact that the magnetic field is weakest at the poles and so drops charged particles in that area. The significant thing about the magnetic field is that it accelerates positively charged protons in one direction and negatively charged electrons in the other direction.

This would mean that not only is the magnetic field useful as a particle accelerator. It can be used to accelerate an orbiting spacecraft by giving the spacecraft structure a negative or positive charge.

IDEA # 294; CLONE COPIER: Eventually one has to wonder. When will we be able to just make anything we want if we can supply all the required atoms?

If we could excite atoms to cause them to emit a bit of radiation, it would conceivably be possible to sense the locations of those atoms in an object. If a sensor device could detect the presence and location of individual atoms and then store this information, it would open the possibility that a load of atoms could then be arranged to make an exact copy of the original object.

The sensor would probably detect atoms one element at a time, starting with hydrogen. The loose atoms could then be put in a chamber of some kind that would use energy to rearrange the atoms into the pattern in the sensor's information storage. Presto, it would take the atoms such as carbon and hydrogen out of any old organic matter and create our favorite meal.

IDEA # 295; SPOT PHOTOCOPIER: Do you know what would make taking notes out of a book a lot easier? A spot photocopier, that's what. It would be great for students to have a wand with a laser at one end that would be scanned over a text one line at a time. The information would be stored on a memory chip in the wand. When the student was done going over the text, the wand would be taken to a printer that would then print out all the text that was scanned with the wand.

IDEA # 296; COMPLETELY SYNTHETIC SODA FLAVORS: Most fruit-flavored soft drinks nowadays are 'artificially flavored' as the label usually reads. You might be drinking grape soda but it probably contains no actual grape juice. The chemicals that give grape it's flavor can be made artificially in order to produce the drink.

Why not just produce entirely artificial flavors? Who says that something has to be grape or orange or cherry soda? Just because these are flavors that people have been used to for centuries is no reason that chemists today have to keep copying them. We should be sure enough of our abilities with chemicals by now to research what really stimulates the human taste buds and creates the feeling of refreshment and then synthesize a soda accordingly without intending it as a copy of natural flavors. Maybe we can do better than nature.

IDEA # 297; CHART EVERY DETAIL OF EARTH DEVELOPMENT TO FIND RESOURCES: This is an idea for after the Human Genome Project is finished. Why not start at tracing every detail of the formation of the earth? How the earth came together from debris in space. How it settled into it's present condition. The bonus of this is that it would show us where the resources are to be found.

IDEA # 298; THE CELESTIAL MERIDIAN AND COORDINATE SYSTEM: Astronomers have compiled extensive star charts to locate celestial bodies. There is a system of stellar coordinates resembling the system of latitude and longitude used on the earth's surface. Planets, which move in the sky along the so-called ecliptic, often have their locations described relative to the background constellations. The planet Mars may be said to be in Gemini or another constellation of the zodiac. The zodiac is the twelve constellations found along the path of the planets through the sky. This is the way planets have been located since ancient times.

Just as positions on the surface of the earth are described relative to the equator, the prime meridian and, the poles, the stellar coordinate system used by astronomers is based at the so-called "first point of Aries". This is where the apparent path of the sun among the stars, known as the ecliptic, crosses the celestial equator, which is a mirror image of the earth's equator in the sky. One hundred eighty degrees opposite the first point of Aries in the sky is the first point of

Libra. In the stellar coordinate system, what we call latitude on earth is called declination and what we call longitude on earth is known as either right ascension or sidereal hour angle.

My opinion is that we need a new system. All of our locator systems for celestial bodies are very old and are designed for observation from earth but not for travel. In such old systems, we care not how to get to a body in space because we could not practically think of doing so until relatively recently. All that the old system is concerned with is describing where a celestial body can be found in the terrestrial sky.

Have you ever stopped to think how difficult it is to chart a specific point in the solar system or elsewhere in space? Space travel, whether manned Apollo missions or unmanned explorer craft, revolves around planets and moons, their positions at a given time and their gravity. We use the planets as reference points but the planets are always moving.

Could anyone easily describe the exact point in space where the Apollo 13 mishap, the explosion of the oxygen tank, occurred? To do so, it would be necessary to use the date method. Locating the position of the earth in it's orbit around the sun, in other words the date when the incident occurred. Since the orbit of the moon around the earth does not coordinate with the orbit of the earth around the sun, it will be necessary to establish the location of the moon relative to the earth at the time the incident occurred. Only then could we describe the point at which the Apollo 13 incident occurred, as long as the distances and angles between the earth, the moon and, the spaceship are known.

In this day and age, thirty-four years after man first stepped on the moon, we still do not have a space coordinate system by which we can readily describe any point in space without relying on celestial objects as reference points. I find this really incredible.

We have had the stellar coordinate system for a long time but it was designed for viewing from earth and not for actual space exploration. Our present space coordinate system tells us where to find the star Arcturus in the night sky, it also tells us where we can find Vega. But it does nothing to tell us how Arcturus and Vega relate to each other. It is an earth-based system designed for observation only. If we are ever going to really explore space, a locator system for exploring rather than just observing will become a necessity.

I believe that the answer is a coordinate system resembling the existent space coordinate system, except that it will be used more like the latitude and longitude system applied to the earth's surface. The space coordinate system will be designed for exploration, not for observation, and thus will be three-dimensional rather than two-dimensional. This system will not only be of great benefit to astronomy and space exploration as well as cosmology but, most importantly, will change the basic mindset concerning space and be of primary importance in getting space exploration back on track.

Such a coordinate system would make easier such tasks as charting the gravitational watershed of a celestial body. We could call gravitational watersheds "gravispheres", meaning a celestial body's gravitational field. This would open the possibility of expressing the gravitational influences at any point in space at a given time. We could also easily describe what the sky looked like from any point in space.

These are tasks that could be done today, albeit with great difficulty. A space coordinate system would make it quick and easy.

Let's begin by reviewing our system of latitude and longitude that we use on earth. The system is based on the fact that the earth is a sphere. Various circles on the earth are defined by the earth's relationship to the sun. The earth rotates on it's axis, which is defined by the north and south poles.

The earth, on it's axis is tilted twenty-three and a half degrees from the perpendicular with the plane of it's orbit around the sun. Thus as the earth goes around the sun, for half the year the northern hemisphere points toward the sun and the southern hemisphere away and vice versa. When your hemisphere (northern or southern) is facing toward the sun, you have summer. When it is facing away, you have winter.

If the earth were a perfect sphere, it would be divided in half by the equator. This is a line going around the earth and is halfway between the point furthest north that the sun is ever overhead and the point furthest south that the sun is ever overhead.

The northern and southern limits that the sun ever appears overhead are known as the tropics and circle the earth twenty-three and a half degrees north and south of the equator. The northern limit is known as the Tropic of Cancer and the southern limit is known as the Tropic of Capricorn.

When the earth is in that position in it's orbit in which it's axis is perpendicular to the plane of it's orbit around the sun, the sun will be overhead at the equator and it will be either the first day of spring, the vernal equinox, or the first day of autumn, the autumnal equinox. The autumnal and vernal equinoxes are where the celestial equator crosses the ecliptic in our stellar coordinate system. Equinox means equal and thus, when the sun is directly overhead at the equator on the first day of spring and the first day of autumn, the length of day and night is equal.

The celestial equator is a mirror of the earth's equator in the sky. It is not exactly the same thing as the ecliptic, which is the apparent path of the sun across the sky in the course of a year. The reason for the difference is that the ecliptic is the apparent path of the sun across the sky in the course of a year and is roughly the plane in which the planets orbit the sun. The earth, however, is tilted the twenty-three and a half degrees from the perpendicular.

The difference between the ecliptic and the celestial equator can be charted as a sine wave. If the celestial equator is the x-axis, the ecliptic starts at zero where it crosses the celestial equator at vernal equinox. Using the sidereal hour angle model of celestial longitude, the ecliptic reaches a maximum of twenty-three and a half degrees south of the celestial equator at ninety degrees, or a quarter of the way around the celestial equator. The ecliptic crosses the celestial equator again at one hundred eighty degrees and reaches a northern maximum of twenty-three and a half degrees at two hundred seventy degrees on the celestial equator. The ecliptic meets the celestial equator again at zero degrees.

When the sun is overhead at the Tropic of Cancer, it will be the first day of summer in the northern hemisphere and the first day of winter in the southern hemisphere and vice versa when the sun is overhead at the Tropic of Capricorn.

Mirroring the tropics are the Arctic Circle and the Antarctic Circle. These circles are twenty-three and a half degrees, in other words sixty-six and a half degrees latitude north and south, from each pole and are the northern and southern limits of where the sun shines every day of the year.

The northern hemisphere is defined as the ninety degrees from the equator to the North Pole. Since the equator goes all around the world, this makes the northern hemisphere one hundred eighty degrees in all. The southern hemisphere follows the same concept from the equator to the South Pole.

The western hemisphere is the one hundred eighty degrees to the west of the prime meridian. The eastern hemisphere is the one hundred eighty degrees east of the prime meridian. Diametrically opposite the prime meridian is the so-called International Date Line.

Latitude is defined as the angular distance on the earth's surface north or south of the equator. Latitude is expressed as north or south and lines of latitude never intersect. The equator is 0 degrees, the poles are 90 degrees north or south. Longitude is defined as the angular distance on the earth's surface east or west of the prime meridian. Longitude is expressed in east or west and lines of longitude meet at the poles and are furthest apart at the equator.

This system revolutionized travel on the earth's surface. Any point on earth could be accurately and easily described. It is difficult to imagine the last four hundred years without this system of latitude and longitude.

As it turned out, the key to using this system was the development of a very accurate clock by England's John Harrison that was rugged enough to be carried around on an ocean-going ship. Such a sea clock could not be based on a pendulum because the pitching and rolling of the ship in rough water may affect the timing of a pendulum.

The clock was set to Greenwich Mean Time, or GMT. Greenwich is a suburb of London and is the location of the British Naval Observatory that determines the exact time by observation of celestial objects. The prime meridian was defined as the line from the North Pole to the South Pole passing through Greenwich. In the system of latitude and longitude, the prime meridian represents 0 degrees longitude.

It was easy to determine one's latitude anywhere on earth. All that is necessary is to measure the angular distance of the North Star (or it's equivalent point in the southern hemisphere) above the horizon. That is your latitude. Niagara Falls is forty-three degrees north, meaning that the North Star will appear forty-three degrees above a flat horizon from Niagara Falls.

Longitude, however, is trickier to determine. The eventual solution was to use the accurate clock on the ship set to GMT. Local solar time was determined on the ship by the position of the sun. The difference in time revealed the longitude of the ship. Since there is twenty-four hours in a day and a complete circle is three hundred sixty degrees, every hour of difference between GMT and local time represented fifteen degrees east or west of the prime meridian. This was a revolutionary breakthrough in navigation.

What I want to do is apply the same concept to space. Not as a two dimensional stellar coordinate system that we have now to describe where an object can be seen in the sky but as a three dimensional coordinate system by which we can describe the actual location of any point in space. Such a system will be necessary sooner or later if we are going to seriously explore space. I am certain that such a system will revolutionize space travel much as the latitude and longitude system revolutionized global travel.

The universe has four dimensions, three spatial and one of time. We should be able to quickly and easily describe the location of any point in space by use of these dimensions. What we need to start with, of course, is a starting point.

In order to describe the location of points on earth, we first had to define the starting point of the system. Such a system is no good if everyone is measuring from a different reference point. The equator was established as the reference point for latitude on earth and the prime meridian was defined as the reference point of longitude.

This is where we encounter difficulty. In space, everything seems to be in relative motion. Our latitude and longitude system on earth would not be of much use if the equator and the prime meridian kept moving. It would seem natural to have the familiar earth as our reference point because it is, after all, from where we are starting. The problem is that the earth keeps moving. Any space coordinate system using earth as a reference point would inevitably involve complex calculations as a result of the earth's movement around the sun.

To get started, I would like to define a straight line from the sun to the star Regulus and extending beyond both as the Celestial Meridian. The center of the sun will be defined as the origin of a three-dimensional space coordinate system.

Thus, my Celestial Meridian is solar-based and not terrestrial based. From space, bright stars are convenient because it makes it easier to fix positions. The line from the sun to Regulus is defined as the x-axis of the new space coordinate system. The y-axis of the new space coordinate system will be defined as perpendicular to the x-axis and in the plane of the earth's orbit around the sun. The z-axis will be defined as that line that is perpendicular to both the x- and y-axes.

I chose Regulus as the marker for the space meridian for several reasons. First it is a well-known star, the bright star in the constellation Leo. Second, it is located right on the ecliptic. Third, it is easy to locate from earth. Many people know that the way to find the North Star is to look for the Big Dipper and follow an imaginary line upward from the two stars on the bowl away from the handle. At about six or seven times the length between those two stars, one comes to another star. That is Polaris, commonly known as the North Star. It has been used to tell directions since ancient times.

To find Regulus, follow these same directions from the bowl of the Big Dipper. However, this time use the opposite two stars in the bowl, those closest to the handle, and go in the opposite direction down from the bowl instead of upward. In about the same distance from the bowl to the North Star, one comes to a bright star. That star is Regulus. The star is seventy-five light years from us.

So there we have it, the basis of our new space coordinate system. Using this system, the motion of any body can be described by formula and the extent of any gravitational field can be charted. Three dimensions, four if time is to be included, and units is all that is necessary to pinpoint any point in space.

Any star or galaxy can be defined by it's coordinates using this system. I believe that it will make the system more convenient to have both the x- and y-axes in the plane of the earth's orbit around the sun. However, it is important to remember that this corresponds to the ecliptic and not to the celestial equator due to the earth's tilt on it's axes. Neither does it correspond to the plane of our galaxy, which is not the same at all as the plane of the earth's orbit around the sun.

To find the distance from any point in the system to any other point, the Pythagorean theorem works in 3D. Trigonometric functions will give the angular position in the sky of any point or celestial body to any other point or body.

This coordinate system is just the beginning. I envision an entire universe of imaginary cubes of any dimension we wish and starting from any point we choose. Coordinate systems can be developed for any purpose, can use any convenient unit and, can start from any convenient point. It is also possible to set up a coordinate system based on the plane of our galaxy, which is shaped like a pinwheel.

All that is necessary when developing new coordinate systems is to describe how the new system, it's origin point, it's axes and it's units, relates to previous systems or to the original system. To create a new coordinate system, it is necessary to decide on a point of origin and an x-axis. Then in which plane the y-axis will be in that intersects the x-axis at the point of origin. The z-axis will then be that line that is perpendicular to both the x- and y-axes.

Systems can be created for specific purposes using polar, as opposed to rectangular, coordinates and a system can be expressed in one or two, as opposed to three dimensions if that is convenient. A coordinate system can be dynamic, or in motion, as opposed to static. Since, in the universe, the definition of motion is relative anyway. The motion of the origin point would be described as part of the system.

The ultimate coordinate system will be based on what I call "point zero", the spot in space where the big bang occurred and thus where the universe started. In this universal system, a line from point zero to the so-called local group of galaxies could be the meridian, or starting point.

In future travel in deep space, navigation will be a throwback to pre-radio navigation on earth. Signals based on electromagnetic waves will not be practical due to the limit of the speed of light. Sightings on stars will determine position. Sightings on two or more stars with known coordinates can be used to determine the observer's coordinates. Just like in the days of sailing ships on earth, the entire sky is a giant array of navigational beacons revealing one's exact location. Stars may look similar but can be differentiated by spectrum.

There is nothing at all alien about this idea. It is very rooted in familiar ideas. The system itself is the same in form to the latitude and longitude system on earth, except that the space coordinate system is three-dimensional and thus measurement will not be in angular degrees. The concept of coordinate subsystems can be compared to a world atlas. The atlas may show where we are in the world but will not help us find our way around a town. For that, we need a map of the town. The spirit of the Metric System, the ready interchangeability of units is mirrored in the fact that any point in a coordinate system can be the origin point in a new coordinate system.

There are two basic types of mathematical coordinate systems; rectangular and polar. Both types can be set up in either two- or three-dimensions. A rectangular coordinate system will have three axes in a three dimensional system while a polar system will have only two.

A rectangular system resembles a street grid and a point will be located by expressing it's location on both axes, or all three axes in a three dimensional system. A polar system expresses the location of a point by it's direction from the origin in angular degrees and it's distance from the origin. A two-dimensional polar coordinate system will have only one expression of angular measure but a three-dimensional system will have two.

In a three-dimensional polar coordinate system, two dimensions will be angular in expression while the other one will be linear in expression. In a rectangular coordinate system, all three dimensions will be linear in expression.

The two systems are, of course, inter-convertible. We can readily translate from one into another. Each has it's advantages and disadvantages. If we can express the location of a point in rectangular, it will have more accuracy at long distances than it would if expressed in polar. Rectangular makes it much easier to determine how far one remote object is from another. If one is located at the origin of a polar system, it will give a much better sense of the distance and direction of the remote point. It will also be considerably easier to express the location of the point in polar.

The two basic types of coordinate system, rectangular and polar, are related to the two great shapes in the world around us. These are the circle (or sphere) and rectangle (or box). The sphere and the circle is of nature, the rectangle and the box are of man. Polar is the coordinate system of the circle, rectangular of the rectangle.

Nature forms spheres by gravity because it requires the least energy to form. A circle forms whenever something spreads out evenly from a point of origin, or a sphere if we are dealing with three dimensions. The universe, emanating from the Big Bang, is presumably a sphere.

Just as the sphere is of nature, the rectangle is of man because it gives the greatest space efficiency. A box does not have the efficiency of space enclosure per volume of material that a sphere has but the great advantage of rectangles or boxes is that they fit together with no waste of space. The only way that spheres can fit together with no wasted space is one inside the other. Thus, this is the shape of houses, bricks, boards, lots, boxes and, suitcases. A rectangle or a box can be said to have much greater organizational efficiency than a sphere.

The human body and animal bodies are combinations of spheres and boxes. The sphere has the lowest surface area per volume and so produces the most compact body but the box has the greatest organizational efficiency.

Right angles would scarcely exist if not for human beings. The circle is overwhelmingly the shape of nature but for organizational efficiency, the rectangle is supreme. Can you imagine if we had houses like spheres and planets like boxes? The first step in organization, other than by the primal force of gravity, is the change from circle to rectangle. Houses and street grids tend to be rectangular although the shape of the town may be circular.

Things of human beings often do incorporate the sphere because it is the most compact shape when interaction with nature is a significant factor. A building would lose less heat in the cold weather if it was spherical but this advantage would be negated in most buildings because of the loss of organizational efficiency.

A V-shaped internal combustion engine has a considerable amount of sphere incorporated into it's design, such as the shape of the cylinders, to accommodate nature since the explosion spreads out in a spherical pattern. The shapes of moving man-made vehicles, like the human body, tend to incorporate the spatial efficiency of the sphere with the organizational efficiency of the rectangle. This can be seen in airplanes especially. Trees also combine the spatial efficiency of the sphere with the organizational efficiency of the rectangle.

In the opinion of this writer, a rectangular coordinate system will be a far superior choice for the charting of the physical location of points in outer space. We are basically using a polar system now in the celestial coordinate system longused by astronomers.

In a rectangular system, everything is related and easily convertible. It is easy to switch from one system to another. It works in the same spirit as the Metric System. Polar is useful for viewing in the same way that rectangular is useful for traveling.

The reason for this is that a polar system is quite difficult to switch from one system at a given point of origin to another system at a different point of origin. A polar system is useful for viewing only if we are at the system's origin, if not then it becomes quite awkward to use. Conversion of one polar coordinate system to another involves breaking the angular lines down into their x and y components anyway, which is the components of a rectangular system.

A rectangular system is easy to switch from one static system to another or to a moving system. This space coordinate system must be a rectangular system. This is not to say that a polar system does not have it's uses, after all we see in polar, our eyes locate an object and look in that direction. This is another example of the use of the spherical when interacting with nature. But for the inter-convertibility and organizational efficiency that we need to become masters of the universe, we must use a rectangular system.

So what I want to do is to switch us from our old polar way of seeing outer space to a new rectangular way. In the polar way, the earth is the point of origin and we think of celestial objects in terms of their angular directions in the sky. I want us to think more in terms of the latitude and longitude system used on earth.

This is very much an issue of dominance and of mindset. When nature dominates, we think in polar. But when man dominates, we think in rectangular. A city built on a grid pattern denotes a well-planned city.

There is no better example of this than the latitude and longitude system. When this system was established, it was saying boldly that the earth is now ours. It effectively belongs to us. We are it's masters. I see the establishment of the terrestrial system of latitude and longitude as a great turning point in the relationship of man to the earth. Even though it would still be a long time before the earth could be considered as well explored.

The coordinate system will naturally require a linear unit of measurement. The system is completely unit-flexible. The most convenient unit for a particular task can be pre-selected by computer. A very natural unit for measurement of the vast distances in inter-stellar space would be the light-year that is in use now. A parsec is also a natural unit of distance appropriate to space. One parsec is equal to 3.26 light-years and is defined as that distance from which the distance between the earth and the sun (93,000,000 miles) would appear as one second of arc, which is a sixtieth of a minute, which is one sixtieth of a degree.

I am a great believer in the use of natural measurement units. A unit should contribute to making a task easier. Units should make calculations and estimations simpler and should make it easier to see things that we may not see otherwise. The absolute scale of temperature is a fine example and it contributes to an understanding of what heat really is.

For closer distances, such as found in the vicinity of a planet or on the surface of the planet, the most convenient unit could be chosen for a one-time task. All we need to know is how exactly it relates to more standard units. Such convenient units for temporary purposes could be called ad-hoc units. For example, we could take the distance that an object will fall in one second near a planet's surface and use that unit there.

Human beings will naturally be less flexible with units of time than with distance. Since we are used to being awake for sixteen hours or so and then sleeping for eight hours. No matter how far we get from earth, we will probably still use earth days as our primary units of time for the indefinite future.

As far as natural units in the universe go, the thing that really counts in the universe on a large scale is the expansion of the universe. We still tend to use light-years or parsecs to describe distances across the universe but I find these to be inconveniently small units. The universe is more than fifteen billion light-years across. Talking about distances across the universe in light-years is akin to describing the distance from London to Los Angeles in millimeters.

In closer, but still in the realm of stars and galaxies, it is light and gravity that counts most. By far, the most logical unit of velocity in the universe is the speed of light since that is the maximum possible velocity and it will be such a natural unit that will reveal what point in space-time could have been in contact by electromagnetic radiation with what other points in space-time.

The space meridian that I have chosen is a straight line from the center of the sun to the center of Regulus and continuing in both directions. We should realize, however, that since the universe is expanding, Regulus and the sun are moving along with our so-called local group of galaxies. This local group is so-called because it moves through space as a gravitational unit and so the galaxies in the local group are not moving apart from each other due to the expansion from the Big Bang.

This local group includes our galaxy, the large Andromeda Galaxy that is about two and a half times larger than ours and, the two smaller galaxies visible from earth's southern hemisphere called the Magellanic Clouds.

Since the rest of the universe outside the local group is moving away from us in the expansion, this will have to be reflected in the coordinates of any point in the universe outside the local group when using our coordinate system.

Alternatively, we could define an entirely new meridian, one that would be more suited to and based on the universe as a whole. I would like to introduce the

Point Zero Meridian. Point Zero is what I like to call the place in the universe where it all began, where the big bang happened. The entire universe was once smaller than an atom and has been expanding outward ever since. The matter of the universe is not expanding through space, more accurately space itself is expanding. Some day, we may be able to pinpoint that spot where it all began, Point Zero.

The Point Zero Meridian will be defined as that line from Point Zero to the center of our galaxy and continuing in both directions. Thus the galaxy, as it moves outward in the expansion of the universe, is moving along the Point Zero Meridian. It is necessary to base this meridian on the center of the galaxy and not of the sun simply because the sun is moving around the center of the galaxy and thus cannot be stationary along an axis from Point Zero.

By the way, our galaxy contains about a hundred billion stars and is only one of about the same number of galaxies in the universe. Our galaxy is unimaginably large, about a hundred thousand light years across. The solar system that we inhabit is about thirty thousand light years from the center of the galaxy. Our sun is just an average star, many of the stars that you can see at night are hundreds of times brighter than our sun.

This Point Zero Meridian will form the x-axis of a coordinate system encompassing the entire universe. We will need to define a y-axis for the system. Let's make the y-axis that line passing through point zero, perpendicular to the x-axis (the Point Zero Meridian) and in the same plane as our galaxy. The galaxy we live in is a so-called pinwheel galaxy, meaning that it is like a flat plate with spiral arms and a bulge in it's center. Of course, we are not yet in a position to take photographs of our own galaxy but you can see photos of the Andromeda Galaxy, which is also a pinwheel galaxy.

The general shape of our galaxy is like a flat plate, meaning that it lies in a certain plane and the y-axis of the point zero coordinate system will be defined as in that plane. Remember that this is not the same plane at all as the orbit of the earth around the sun. The z-axis can then simply be defined as that line that is perpendicular to both the x- and y-axes.

There are two possible types of point zero coordinate system, one that expands with the universe and one that does not. With a fixed system, one that does not expand with the universe, more points will be continuously brought into the system. Since the universe is continuously expanding, the Point Zero System will have points that are existent or non-existent with regard to space and time. Every point is either existent or pre-existent on the system (since it looks now as if the universe will keep expanding forever).

Possibly we can begin charting what I will call "zero units". A zero unit could also be called a "gravitational cell". That is a unit such as our local group of galaxies that moves as a unit in the expansion of the universe. Revolutions and rotations of celestial bodies, including that of our galaxy can be called "local gravitational circles" and such motions are not included in the expansion of the universe. An expanding, as opposed to static, point zero coordinate system may make it easier to sort out and chart the zero units in the universe.

For now, the most important effects of this space meridian will be a change in mindset with regards to space. As far as I am concerned, space exploration should be much further along than it is now. I remember the summer of 1969, it was my first summer in America and I was eight years old, almost nine. There was that incredible day when men first walked on the moon. It seemed that anything was possible by the time I grew up.

Except that it did not happen. At least not like it seemed that it would. The manned space program faltered, after America had beaten the Soviet Union to the moon part of the reason for wanting to go there was gone.

One of the reasons for this Celestial Meridian is to change our mindset about space exploration. Space is not just something to be observed from earth, space is an extension of the earth.

Public backing is vital for space exploration. Part of the problem is how we treat space. We act like it is there only to be observed. It is not really "ours" like the surface of the earth is. The leading nations in technology are not led by powerful kings that can compel sailing expeditions without consulting the public as was done in the days of exploration centuries ago. Today, outer space has to compete with cyberspace for people's attention.

My Celestial Meridian is to make space into our turf, in the same way as the surface of the earth.

In space, not only should we be thinking outside the box, we should be thinking outside the whole earth. In many ways, we still have the mindset that earth is the center of the universe. To really get on with space exploration, I believe that we must break this earth-centered mindset. Our earth-centered mindset is limiting us in ways that we do not realize.

Our earth is not suitable to serve as the origin point of a space coordinate system for the simple fact that it is continuously moving around the sun. At the distance of stars, this would not be as true because the distance between the earth and the sun would be insignificant. I consider it beneficial that the origin point of my space coordinate system is the center of the sun rather than the earth. This will help to break us out of this earth-centered mindset.

I can think of no better way to break out of our confining earth-centered mindset and get space exploration back on track than to "mark our territory" in space with a coordinate system based around the Celestial Meridian that relates to space as a three-dimensional entity to be explored rather than as a two-dimensional entity to be merely observed.

If astronomy were developing today, we would have thought of this already. Astronomy, however, is the oldest of the sciences and is very bound by tradition. The celestial bodies could be observed for as long as people have been on earth but it is only relatively recently that we could think of actual travel in space or even knew what the real nature of the universe is.

There is thousands of years of tradition of charting a celestial body by it's apparent position as seen from earth and not it's actual position. We are in the twenty-first century but we still think in terms of constellations. We have millennia of experience in thinking of line of sight with regard to objects in space. Mars is not in Gemini. It never has been and never will be in Gemini. The constellation Gemini does not exist except as an apparent grouping of stars as seen in line of sight perspective from earth. Gemini's stars are actually many light-years distant from each other.

Space was always something that we could see but not reach. Like the crown jewels or a famous painting in the Louvre. The universe is awesome but we consider it as awesome in the way that we can see but not touch because the distances are too great. My Celestial Meridian and Coordinate System is to move our mindset in a new direction.

To be sure, space is very different than earth. We live on the surface of a sphere. When we look at space, we look up at the apparent inside of a hollow sphere. On earth, we cannot see distant places but we can travel there. In space, we could see distant places but could not travel there. Astronomy is not usually a science that we can get our hands on like the other sciences. Much of the physics of space is not found on earth.

The result is that we do not really have the mindset that is required for the enormous undertaking of space exploration.

The Celestial Meridian and three-dimensional coordinate system is an approach to encourage a new mindset toward space using historical patterns. This approach is in no way new and unfamiliar but is deeply rooted in our history.

This is about not only thinking outside the box but, thinking outside the entire earth. It is a new way of thinking about space. With the implementation of this system, the universe is our turf. It is not just something to ponder in awe. Giving people the coordinates of something makes it seem much more accessible.

Think of space not as the ceiling of a cathedral to be looked up at but as the floor of the cathedral as we think of the earth.

When we believe that the universe is really ours, fact and feeling will feed on each other in a constructive spiral. This is certainly no breakthrough mathematics, just a new way of thinking. Just having this system will change our attitude toward the universe. This new mindset will help not just with space exploration but with astronomy and cosmology as well.

The mindset changes that we have previously gone through with regard to the universe have shown us how insignificant we are. The realization that the earth revolved around the sun and not vice-versa, the discovery that we live in an enormous galaxy but that it was only one of about a hundred billion made humans on this little earth feel like ants. It is time for a shift in space confidence. From now on, we are the masters. If there are any breakthroughs ahead in space travel, this will help to bring them out.

There are amazing parallels between early explorers and space exploration today. The orbiting of the earth parallels the sailing ships following the coast of Africa instead of venturing out beyond sight of land. The early trans-Atlantic voyages can be compared to the Apollo moon missions, astronauts brought back moon rocks just as the explorers brought back new world souvenirs.

Sailors in those days did have the advantage over us that their journey was not quite as much of a technical challenge and powerful kings did not have to consult the people. Today in the space age, we have the advantage that we can see the places we want to go and can explore via electromagnetic waves long before undertaking the journey. The mindset of exploration in those days was much assisted by the crude maps and the new system of latitude and longitude. This system gave people the impression that the world, even it's furthest corners, is "ours".

Thinking that space is really ours is largely self-fulfilling. It is similar to the new system of latitude and longitude in the days of sailing ships. The reason that space exploration is progressing nothing like it could be is the age-old way that space is presented to the public, as something to be observed but not to be physically explored.

Usually, in the course of exploration, we nail something down and then map or categorize it. But we can use this in reverse. Early maps of the world were very crude and inaccurate but such maps spurred men to discovery. This is using the same approach to space.

Consider the effect that the latitude and longitude system has had on the exploration of the earth's polar regions. It had been known for a long time that the earth was spherical and was turning. But it was not until the system of latitude and longitude was developed that explorers began the quest to reach the poles.

The system of latitude and longitude had the same effect on world exploration, not only in a navigational sense. When it was developed it made it seem as if this great big world is really ours. If we can define a point on the surface of the earth, we should at least know that it is there. Before we can find a way to do something, we must first conceive of doing it. When early explorers saw maps of distant places, however inaccurate, and those places categorized by latitude and longitude, they somehow found a way to get there and in doing so, did much to bring about our modern world.

So it is with space and this system of the Celestial Meridian and Coordinate System.

It is very important that we have a three-dimensional coordinate system. Our mindset is not only way too earth-centered. It is also very two-dimensional. The third dimension is what we can call "the missing dimension". In the space age, just about everything is weightless and three-dimensional. Until the twentieth century brought us aircraft, spacecraft, skyscrapers and, submarines, human beings lived in a very two-dimensional world.

On earth, we have been limited by gravity and do not even realize how limited we are to two dimensions. Aside from sports on a playing field, games like chess, checkers and, board games have been played in only two dimensions. To truly develop the required mindset for space exploration, we not only need to think of the universe as ours, we must get much more accustomed to thinking in three dimensions.

I believe that this system of the Celestial Meridian and three-dimensional coordinate system is the first step. In such a chart, color-coding or shading can be used to indicate depth. Alternatively, two charts as seen from a right angle can be used. Of course, everything can be computerized and a "living map" can show the movement of bodies in relation to the passage of time.

IDEA # 299; THE THEORY OF PARABLES AND THE GREAT EXPERIMENT: Jesus often used parables to explain religious concepts to his audience and apostles. Parables were also used by other biblical figures such as Paul. My theory is that these parables were only a sample. God has designed the entire world in which human beings live their lives to nudge us in his direction. There are unwritten, unspoken parables everywhere.

This is another fine example of seeing things in patterns. The patterns in the world all around us are to introduce us to the patterns of God. Even though a natural human being will move away from God rather than toward him, he has created the world so that the patterns that occur will counter our natural ungodly tendencies and nudge us toward him.

The purpose of the human race, as created by God, is an experiment. An experiment is the asking of a question of reality. God created humans after Satan rebelled against him. God wants to prove through the experiences of the human race on earth that his ways are right and Satan's ways are wrong. At the same time, he wants to test each and every one of us to determine who should be permitted to enter his kingdom.

To prove that God's ways are right, this great experiment is very far-reaching. The world is made so that human beings encounter and deal with a vast range of issues and situations. God's ways must be proven right before he destroys Satan for rebelling against him and there is a wide range of ideas to be proven wrong. We probably do not begin to realize all of the false concepts in the spiritual realm being proven wrong by our existences and experiences.

Human beings, in this great trial, not only encounter a very wide span of issues and situations. People face the world with a wide variety of educations, hereditary factors, habits, temperaments and, skills. Life takes place in a broad range of natural, technological and cultural environments.

To accomplish God's purpose for us, we had to have a certain level of intelligence and sophistication. Animals must depend on inbred instinct to survive and that makes them unsuitable for the purpose that God has for humans. A sophisticated creature with free will and capable of building civilizations was what God wanted and that, of course, is where we come in.

It is an adequate but imperfect world and race of humans that is required to show God's truth. If we were perfect or lived in a perfect world, we could not prove God. The simple fact is that a perfect situation is not the kind of test that God had in mind.

In our world, you may have noticed that things do not work flawlessly. Those in authority make errors and there are such things as poverty and wars. Our knowledge is adequate but imperfect. Our wants, our needs and, our possessions may all be different. We would not understand right if everything was right and if all was well, we probably would not have learned to appreciate it anyway.

If the Word of God was not against the grain of human nature, humanity could not prove as effectively that the ways of God are right. Life gives us the chance to sin or not to sin and at the same time shows us the results of sin. Generally, doing right instead of wrong brings us a good life but not always or else we would do good things just for our own benefit instead of out of loyalty to God.

My Theory of Parables is that the parables used by Jesus and other biblical figures are just a few examples of all possible parables. Christianity is certainly not what we would call a "nature religion". But the world all around us, virtually everything we deal with, is like God's open arms beckoning us in his direction.

God gave us free will and will never force us to accept him and an understanding of the Bible is essential. But while the Bible teaches that "the world" is opposite to the ways of God, he has set things up so that there are "signposts" all around us in the form of basic patterns pointing us in his direction and giving us a foundation to understand the principles of the Bible.

Let's start with the human body. First of all, your body could not possibly have come into existence without God. The idea of even the simplest life coming into existence by random collisions of atoms is ludicrous. Sleep, followed by awakening, shows us how death is followed by resurrection. Just as bodily wastes must be eliminated, we must also eliminate ungodly thought and characteristics.

We have hunger to let us know that we need food and our food must be safe and nutritious. Our water must be pure and not salty. In the same way, God designed us with an inner yearning for him (some Christians call it a God-shaped vacuum) but we must be sure that we are getting the correct and pure God in the same way we would handle our diet.

We take regular baths or showers and in the same way we must regularly cleanse our souls. If we had life too easy, we would make little spiritual progress. This is the same as our minds and bodies that need challenges to grow. Exertion makes us sore and tired but we emerge stronger than before.

Physical pain tells us that something is wrong. A wound will get infected and an injury will get worse if not attended to. The pain caused by sin follows exactly the same pattern. We have disease to show us that sin is a disease. In fact, cancer and other infectious diseases imitate sin. Just like cancer, sin either gets cut out or it destroys. Finally, sickness in the here and now helps us to focus on a day when there will be no sickness.

Think of a person as a finite mirror of an infinite God. It is an adequate but imperfect people and world that is required to show God's truth. Other people show us life beyond ourselves. Seeing people that lack what we have or have what we lack gives us a three-dimensional view of life. We see human nature to remind us why Jesus died for us.

Human relationships show us how we must feel to God. There is a big difference between knowing someone and loving them. So it is with God and each of us. Merely knowing about God is not enough. We cannot be faithful without the chance to be unfaithful. This is a model to help us to understand our relationship to God.

On earth, God allows us to have false gods and idols just so we can see them fall. Heroes fail to remind us that everyone is human. God wants us to keep things in perspective. The things of men are not equal to the things of God. Political scandals show us that our leaders are not gods. Even great people show room for improvement.

Through our relationships with other people, God allows us to see the folly of misplaced loyalty. This is to make us careful to put our worship in the right place, the one true God. Strangers show us how we may think that we know someone but really did not. So it is with false gods, messiahs, religions and, philosophies.

A person is a finite mirror of an infinite God. The life cycle that we live points us to God. Our lives moving through all the different stages are a part of the great experiment. The lives of animals do not have enough range to show or not to show the righteousness of God's ways. Watching a birth gives us an analogy to the hereafter and at the same time, the birth process also shows that life cannot create itself.

Family resemblances remind us how Jesus resembles God. Children remind us that we are still children in God's eyes. A father watching over his children is a demonstration of how God watches over, and sometimes chastises, his children.

Childhood gives us an early opportunity to learn about God before we become jaded and hardened. A child in a classroom reminds us that all of life is a classroom. Just as school and apprenticeships prepare us for the future, life prepares us for our eternal future.

Children watch grown-ups handling the world but understand little even when an adult attempts to explain things, so it is with God and us. Children show us how some things are simply beyond us and must be brought to our level. Childhood moves on to adulthood to give us an analogy that we will also be moving on from life on earth. Children move on from toys and games to adult things in a way similar to human beings moving on to God's Kingdom.

Youth also gives most of us memories of carefree times and a near-flawless body to give us an idea of what we may have to look forward to. Aging forces us to reckon with the fact that we will not be here forever and hopefully to think about what comes after this life. The deaths of those around us, particularly sudden deaths, reminds of this yet again. God wants us to keep the big picture in mind.

Throughout life, failure reminds us that we are not gods. Failures keep us humble, which is the way that God wants us to be. Mistakes are not a waste but are very valuable to show us our limitations. Surprises happen to let us see how little we really know.

Many of the parables that Jesus used revolved around nature. The sky represents a glimpse of eternity. No matter what happens on earth, the timeless cycles of days and years and the unreachable but ever-present stars help us to keep things in perspective and not start thinking of the world as the be all and end all. Some trees reach higher than others, the straightest trees grow the highest, but all fail to reach the sky. So it is with people.

The unchanging cycles of days and years shows us our dependence on what we cannot control. We can either be in the spiritual light or in the darkness just as there is night and day. Yet, each day is a new beginning.

The sun and moon show us the difference between the real light, which represents God, and merely reflected light, which represents God's children. The moon, as seen from earth, waxes and wanes in the amount of light it reflects to earth, just as many people do in their walk with God. The night and the clouds remove the sun from our sight but we know that it still exists, so it is with God during life's trials.

Weather and storms show us that the forces of nature do not build complex things. Rather, natural forces tend to break the complex down into the simple. This reminds us that we did not just come to exist, we were created by God. Storms and calm reflects the conditions in the human soul. Rain displays the contrast between what is fun versus what is good for you, few people want it to rain on a nice day but we would have no crops without it. Looking at the clouds and the ground remind us that some things are permanent like the ground, such as the things of God, while some are temporary like the clouds, such as the things of the world.

Different terrain shows us different life perspectives. Looking at things from flat ground or from a hill or from a mountain lead us to reflect on life from certain points. Hills and mountains demonstrate that through effort, we can rise above our natural state but we still fall far short of the stars. Deserts and the arctic, as well as the moon, made us appreciate the verdant earth that God gave us.

The laws of nature are unalterable, so it is with God's laws. The decay of dead matter shows us that nature does not create complex forms from simple ones but instead breaks the complex down into the simple. God must have created life, it could not have come into existence by itself.

Gravity causes things to fall. This is a model to help us understand the fall of the human race in the time of Adam. Water, like God, exists in three forms but is still the same thing. Ice, liquid water and, water vapor corresponds to The Father, The Son and, The Holy Spirit. Water, like fire, can be either a blessing or a curse, depending on how it is handled. Natural disasters occasionally remind us of how powerless and dependent we really are.

Creatures of various levels help us to see what our position is in the grand scheme of things. Animals with instinct help us to understand what free will is. Ecdysis gives us a model for shedding the old ways in favor of the new. Predators give us a living representation of what Satan is.

Daily life includes various tests to help us to understand just what life is. Life does not just continue, it must be maintained with care and nourishment. So it is with our spiritual lives. Spilled liquids or a dropped piece of pottery while cooking show us another example of entropy. The forces of nature tend to break the orderly down into the disorderly. So, life did not happen by itself.

Some places are clean and some are not. Human souls are the same way. Dirty areas tend to breed germs and disease, so it is with sin. Washing shows us how dirt can be cleansed.

Working to make or build something helps us to appreciate God's creation. The learning of a skill is a model for growing in one's Christian life. In employment, authority figures telling us we have work to do mirrors God's Word telling us that we have work to do on our lives. Promotions mean power but also more responsibility, so it is with a growing relationship with God.

There is no better model for the relationship between God, man and, Satan than the one Jesus used; the shepherd, the sheep and, the wolf. Tools show us how such tasks as chiseling, sanding and, polishing work. This shows us how God works on His Children. The refining or ores and flour also display God's workings with us. Arts and crafts give us an example of creator and creation.

Daily dealings also enable us to see the results of dishonesty. Earthly justice gives us an imperfect model of heavenly justice. Fishing shows us how Satan is trying to catch us through lures and temptations. Mining illustrates the search for God and the spiritual truth in the Bible.

Cooking is timeless and it also shows how God works with us. The cooking of food as well as the tempering of metals shows us the value of trials. We use what is useful, the rest goes in the trash fire and so it is with God and his creation. A cooking fire must be stoked to keep going, so it is with our spirits.

Vessels and baskets show us how it is not so much the vessel that is valuable but what is in it. So it is with people, it is what is inside that God cares about.

Someone who is too proud and self-sufficient is of little use to God just as a grape must be crushed to make useful wine and wheat broken to provide bread.

Buildings are also a timeless part of human existence. If you walk past a boulder and then walk past a house, you will see how a house could not possibly have built itself. So it is with a person, millions of times as complex as a house.

Buildings are an attempt to tame nature and according to the Bible, our own natures must also be tamed. A building protects us from the elements but to benefit from this protection, we must stay within the walls. This reminds us to stay within God's laws.

A building will deteriorate if not maintained properly and so will our souls. Buildings can be made of a variety of materials and the durability and reliability of a building depends on what it is made of, so it is with the human spirit. Fences and walls are built to allow us to control what comes in but also remind us to guard what we allow in our minds, hearts and, communities.

Putting oneself in an unstable position high on a building invites a fall. Just as putting oneself in an unstable position in life invites a fall. Heights show us that the more authority a person has, the more damaging is a fall. The higher we climb in the world, the more destructive a fall will be.

Bridges and roofs require support. In the same way, we need "pillars" in the form of not only food and sleep but of time with God. The horizontal beams can be compared to the world in the same way that the vertical beams can be compared to God.

Life often resembles a journey. Any kind of journey can be a model for a spiritual journey. We will get much further going down a straight road in the right direction than with dissipated wanderings. Just as there are wrong turns on roads, there is also in life. Pathways remind us that we tend to go where trails have already been blazed.

Travel should make us think about our directions. Are we getting lost or are we going in the wrong direction? Just as we use navigation and landmarks when traveling, we also should use God's Word in the journey through life. Navigation by stars demonstrates the value of a reference point beyond the earth to guide us.

Different landscapes show that some are much easier to cross than others but all are part of the same earth just as good and bad times are part of the same life. The existence of different languages and cultures show us that we must be prepared when we go to a different place and so it is with heaven.

When traveling, hills give a higher and wider view of things and mountains show a panoramic or "big picture" view. This reminds that we should occasionally stop and look at the big picture of what life is all about. Seeing things from a different point of view may help our understanding but it still cannot change a desert into an oasis. In the same way, it will require more than just looking at life to change it for the better. Distances show us that moving closer to or further away from something also affects our proximity to other things. Life works that way too.

Land and bodies of water reminds us that there are some places that we can walk and some places that we cannot. Our lives must be governed by God's rules, these rules give us places that we are free to walk but some that we are not. The crossing of a body of water or a canyon reminds us that there is a gap that must be bridged between what we are and what we should be. A harbor reminds us that God needs an unobstructed spiritual opening to deliver blessings during our lives. Dangerous currents in water should remind us to beware of the ways of the world around us.

Since the beginnings of civilization, farming has been a paramount occupation of human beings. Many of Jesus' parables involved farming and as far as doing good goes, we reap what we sow just as the farmer reaps crops after sowing seeds. In another parable, Jesus referred to the fact that a plant will not last without sufficiently deep roots and so it is with us in our walk with God.

The progress of crops from the planting of the seed to the growth to the reaping of the fruit or grain mirrors the growth of the Holy Spirit in a believer. A fruit tree shows that the fruit only comes when the roots go deep enough. Plants must be tended and protected from pests just as our growing relationships with God must be tended and protected.

Just as a plant cannot grow without the sun no matter how well it is tended, we need the spiritual light from God. When God plants his seeds, what kind of fruit and how well it grows will depend upon what type of soil the seeds land on. The "soil" is, of course, people.

Rain shows us that less pleasant things are also necessary for growth. We consider rain as a nuisance but plants cannot grow without it. In the same way, we will probably not make the most spiritual progress if we have things too much our own way in life. Pruning certain plants reminds us that sometimes it requires that something be cut out of our lives to make us grow in the direction that God wants us to and ultimately we will bring forth more fruit that if the pruning had not taken place.

There are few better parables than the agricultural ones, which is why so many of Jesus' parables come from farming. The crops and weeds, the poor soil and rich soil, the sowing and the reaping illustrate very well how the Kingdom of God works. That is why God designed things as he did.

Money, in many forms, has also been with us since the beginning of civilization and was also used by Jesus for parables. Gold and diamonds are considered as precious because of how they handle reflected light. So it is with the believers reflecting the light of God. Money gets us used to the ideas of debt and payment, which we need to understand how the Kingdom of God works. Items carry price tags and this gets us used to the fact that actions have price tags too.

Money helps to show us personal responsibility. Investing can be done wisely or not wisely. Some people are rich and some are poor, this shows us that our souls can also be rich or poor toward God.

The progress of civilization from it's beginnings revolves, of course, around the development of knowledge and technology. The mastery of fire was an early step toward civilization. Fire shows that the closer you get, the more warmth it provides. It also shows that something beneficial can be destructive if used wrongly.

Craftsmen make us think what a great craftsman the creator must be. The carving, sanding, polishing and, smelting show us how God works with us. If a piece of metal had feelings, it probably would not enjoy being sanded and polished but this is what makes it into a useful tool or implement.

Combat and armor has always illustrated how the devil seeks openings and opportunities to destroy us. A boat is in the water but the water is kept out. Similarly, Christians live in the world but the ungodly ways of the world must be kept out. If we fail to do this, we "sink" just as the boat does when the water pours in.

Just as people since early civilization have made maps to find the way around, we have the Word of God to find our way through life. Glass windows show us how God can see through any wall or barrier whether or not it is made of glass.

The machines that human beings have constructed show us how foolish it is to suppose that a complex machine constructed itself or appeared by random chance. How much more foolish it would be to suppose that an infinitely more complex human being had come into existence without a creator.

Robots let us see that we are not robots. We have free will and the ability to do what we want. God did not create us as androids. He needed an advanced species with free will to show that his ways are the right ways. Can you imagine a machine that could appreciate beauty and feel a wide range of emotions? That lets us see the wonder of God's creation in comparison to the machines that we manufacture.

Many of Jesus' parables had to do with agriculture. Today, farming has largely given way to manufacturing and information technology as a primary occupation. Yet, these modern fields offer parables to illustrate the Kingdom of God just as agriculture did in the time of Jesus. If Jesus were here bodily today, he would use parables with a modern theme.

Investing and developing of new products shows the many false starts necessary to come to the real truth and it is the same way with religion. Products show us that the may break if they are not used as intended. In the same way, we must live our lives according to God's instructions. Products come with warranties, which are only valid if the products have been used correctly.

Our modern technology and knowledge can be a double-edged sword because it is a product of human beings that have both good and evil in their natures. We have cars and televisions but we also have bombs and guns. Radio shows us how we have to tune in to God to hear his message. We will not hear the message of God when our hearts are tuned into worldly things. A turntable with the needle stuck on a record lets us see how we can sometimes use a jolt in our lives to set us on the right path. Photography illustrates how deeds are not forgotten. God has a "photographic" record of every deed ever done.

We get physical power from tapping into fundamental forces in some way. In a similar manner, we get spiritual power from tapping into God. Electrical switches demonstrate that we must close the switch to get electrical power or we will not receive it no matter how much power is available to us. So it is when we form a connection with God. Batteries provide power but must be periodically recharged or replaced, so it is with our "spiritual batteries".

A car and driver is a good analogy to the workings of the body and the spirit. The spirit "drives" the body in the same way that the driver drives the car. A driver can get out of the car and leave the car "dead" in a way similar to the spirit leaving the body. A car also illustrates that a human being is even much less likely to just come into existence without a creator than is a car.

As with life, driving conditions may be good or bad and distractions can prove disastrous. Some have better cars and more favorable roads than others. But a modest car going in the right direction will end up better off than a fancy car going in the wrong direction. It is the same with human beings and God.

The rules of driving resemble the rules of life in many ways. If a driver ignores warning signs and meets disaster, we would say that while it is unfortunate, it is also his own fault. It is the same for those who ignore the warnings in the Word of God.

An airplane cannot take off for the sky as it was intended if it is too weighed down. To climb above storm clouds into the wild blue yonder above, the aircraft must not be overloaded. Neither are we to be weighed down by the things of the world in our dealings with God.

Another down side to modern technology is pollution of the ground, the air and, the water. Yet, this also illustrates the Kingdom of God. Pollution bears a strong resemblance to sin. It shows how we cannot just bury sin or think that it is gone and forgotten without paying the consequences sooner or later. The world and it's sin has a lot in common with the Love Canal. Just as pollution ruins the natural environment, sin ruins lives.

The interesting thing about scientific discoveries is that they have shown us how extraordinarily intricate God's creation is and how insignificant we are. Around the beginning of the Twentieth Century, humans were gaining evermore knowledge and were beginning to feel like the masters of all there is.

But then it was found how vast the universe really is, that our enormous galaxy is only one of a hundred billion galaxies and that so many of the nebulae visible from earth were not clouds of dust and gas in space as had been supposed but vast galaxies similar in size to our own. It was quite a humbling awakening.

At the same time, it was being discovered how infinitesimal but extremely intricate were the particles that made up the atoms. It was soon to be found how stunningly complex was the structure of even the simplest forms of life. We saw geological processes on earth and the lifespan of the entire universe measured in billions of years. It certainly showed us a glimpse of eternity compared to our brief life spans and human history.

We are but humble observers of all that God has created. I think it can be safely said that God has allowed us to use our advancing scientific knowledge to "put us in our place", to let us see the way things really are.

The ways of human reasoning are not the same as God's reasoning. This is what the Bible repeatedly tells us. One of the best ways to illustrate this is the fact that the earth orbits the sun and not vice versa. To our observation it appears to be the other way around, the sun seems to orbit the earth. But the fact is, it is not the way our reasoning tells us, the earth goes around the sun.

Science shows us that the foundation of the universe is a few laws that are absolute and invariable. It is the same way with the Kingdom of God. Mathematics lets us see how nothing finite can sum or product to infinity. So it is with God.

No matter how much a human being may have going for him, he can never reach God without Jesus. Just how orderly the universe is demonstrates that it must have a creator. When we look at the desolate moon and other planets with extremely hostile environments, we are led to appreciate the earth that God has given us.

The tests done in science laboratories give us an analogy to the vast spiritual test that is taking place on planet earth. The difficulty of predicting weather more than a few days in advance makes us appreciate the perfectly accurate Bible prophecies made thousands of years in advance.

Jesus and God is the same thing and yet are in two different forms. This can be seen reflected in the universe that God created. Electromagnetic waves, such as light, can be seen to exist as either waves or particles. Matter and energy are interchangeable. The same concept applies to state of matter, water can exist in three forms; ice, liquid water and, water vapor. In a similar way, God exists as the Father, The Son and, The Holy Spirit. Electromagnetic waves also remind us that God sees and hears prayers across a distance.

The great knowledge and technology that human beings have gained is an important part of God's great experiment. It shows us what a double-edged sword our own natures are. We have developed wonderful medicines but have also developed fearsome bombs. Technology has certainly not solved all human problems but has created other problems. Our knowledge and technology has failed to prevent many disasters and God allows it to be this way so that we keep our perspective on reality and do not develop undue confidence in our own abilities.

Part of God's great experiment is false ideas. God allows us to pursue man-made utopias to let us see how fraught with difficulty it is. Unjust and repressive governments are allowed to exist to cause us to put our confidence in god rather than in man. Destructive ideologies like Nazism and Communism had to be allowed to exist to show the folly of false paths and that there is a lot more wisdom in the Word of God than in the doctrines of men.

The overall course of human history gives us a light on God's great experiment. The ancients got civilization going. God's Word could not be tested as well in a primitive society. Europe became established as the primary base for Christianity in the world. However, the Moslem World "held the torch" for Europe as the leader of progress while Europe was going through the Dark Ages. The printing press was invented just in time for the Protestant Reformation and I am certain that this was God's doing.

The continuous failure of utopias down through history keeps reminding us that only the Ways of God can result in a utopia. The French Revolution, with it's deep hostility to the church, led not to freedom under human reason but to dictatorship and then war under Napoleon. Robespierre's reign of terror was a terrifying twist on the high utopian ideals of 1789.

The discovery and settlement of the new world gave humanity a chance to do everything over and do it the right way this time. However, it was not to be. People brought their sinful natures to the new world as well. The last hundred years have brought almost unimaginable progress and prosperity relative to all the rest of human history but, true to human nature, it has also brought devastating wars and dictatorships. Dictatorships are allowed to happen to remind us that the Ways of God are much more fair and much less cruel than the ways of man.

The fact that the Twentieth Century was dismal in so many ways goes to show us that there is no salvation in knowledge, technology or, materialism. Humans accomplished so much to lengthen lifespan and cure disease even while developing terrible weapons of destruction.

The course of history keeps reminding us to be careful of where we put our hopes. The utopian ideas early in the last century ended in communism. The "Roaring Twenties" ended in the stock market crash. After Darwin's Theory of Evolution became widely popular, the optimistic secular humanism came crashing down in World War One. The trenches of World War One reminded a world that had forgotten about hell what hell might be like.

The Titanic was supposed to be unsinkable. The Asch Building was supposed to be fireproof. Hitler exclaimed "the Sixth Army could storm the heavens". None of it turned out to be true.

Every time the human race takes a leap of faith in it's own reason, a scientific breakthrough, such as the discovery of the true vastness of the universe, reminds us how infinitesimal we are. No matter what we accomplish, people have only a brief time on this little earth. The earth is only a speck in all of space and this should remind us that our lifespan if just as brief compared to all of eternity.

Whenever we put faith in something other than God, he may show us how unworthy it is of our highest faith. The Civil War showed America how confrontational democracy can be. The depression put a damper on capitalism and materialism. The dissatisfaction during the prosperous sixties reminded the next generation of the folly of materialism. Woodstock reminds me of the multitudes of people coming to hear Jesus preach the Sermon on the Mount. Yet, the idealistic Woodstock counterculture that arose was not based on Jesus and was dampened by the violent anti-Woodstock at Altamont Speedway. The sixties drug scene at Haight-Ashbury turned ugly.

No matter how "logical" it may seem, progressing in a way contrary to God's Word never works out very well in the long run. The Dr. Spock movement toward "freedom" for children led to a very self-centered generation. Modern North America and Europe have unprecedented prosperity but the spiritual emptiness of materialism has led to unprecedented drug abuse. The so-called "sexual revolution" has now died of AIDS. The degree of alienation among younger people and the number of school shootings has given us warning that there is no salvation in prosperity.

Sometimes, God may give warnings on a national level. The devastating 1923 Tokyo Earthquake and resulting fire warned Japan about what it could expect if it went forward with war. The inferno of the Hindenburg gave Germany a glimpse of what it could look forward to in war. The "War of the Worlds" episode of 1938 preceding World War Two let us know that it is not space aliens that we should be worried about but our fellow humans and ourselves. The sniper assassination of JFK and the Bay of Pigs fiasco warned America about the consequences of poorly planned involvement in Vietnam. The 9-11 attacks remind us that we are quite vulnerable and certainly not immune to the devastation of Armageddon foretold in the Bible. Finally, the Washington D.C. sniper warned us what U.S. soldiers could expect in a hasty, poorly planned invasion of Iraq.

If we look at the big picture, we can see how the entire course of world history is a part of God's great testing. The Roman Empire made it possible for Christianity to spread rapidly. If the empire had not existed, it would have been much more difficult for the Word of God to spread as it did. The new world could be considered as a "prize", it was the Christian nations, rather than those of any other religion that got to claim the new world. I believe that this was to show the world which religion was the truth.

I also believe that God greatly supported the Reformation. Not only did the printing press come into use at just the right time to spread the Reformation, the discovery that the earth went around the sun damaged the credibility of the Catholic Church at just the right time to open a wide door to the Reformation. Since the church had always insisted that the sun went around the earth. Furthermore, the bubonic plague had killed priests as easily as anyone and this reminded everyone that no human organization or church "owned" God.

The colonial empires of the European nations served the same purpose as the Roman Empire, enabling missionaries to bring the Word to the world. The colonies would be led to wonder why their god or gods could not deliver them from control of the Christian European nations. However, Europe's diminished role in the world through the Twentieth Century displayed how it had been weakened by it's growing secularism.

I believe that the other religions are largely an attempt by Satan to lead people away from Salvation in Jesus. Let's face it, no more than one religion can be the truth simply because the religions are so different from each other.

Judaism is the oldest surviving religion. Around the time that Judaism began, Hinduism also began in another part of the world. Hinduism, from which the other eastern religions come, is the diametric opposite of Judaism. The fulfillment of Judas'was the coming of the foretold messiah, Jesus. This brought Christianity into existence as the fulfillment of Judaism.

Islam, another monotheistic religion, was later founded and it was claimed that the original intention of the Bible had been distorted. However, I have found that the Bible is the book with all the evidence in it's favor. It fits perfectly with modern science and has many fulfilled prophecies to it's credit.

Evil must be permitted on earth as part of God's plan. If we did have everything perfect, we would not appreciate it. Worse and better things give us a glimpse of heaven and hell. If we were perfect or lived in a perfect world, we would also not be able to prove that God's ways are the right ways as he wants us to.

Part of God's plan is imperfect beings living in an imperfect world. Disappointments and mistakes are things that God has intended us to deal with as part of the great experiment. We also are destined to see clearly that there is no salvation in the things of the world. It often takes a shaking of our faith in the world to bring us to faith in the God that is beyond the world.

Suffering here makes us appreciate paradise and death makes us appreciate life. Handicaps in others should make us appreciate what we have. When is the last time that you thanked God that you have eyes and hands? Worldly justice systems give us a model of what God's justice system is like and injustice in the world leads us to appreciate God's perfect justice.

Part of God's test is for us to cope with human natures that are opposite to God's laws. What we want to do and what is right are often two very different things. That is the way it was meant to be. Life gives us a chance to sin or not to sin and it shows us the results of sin.

To serve it's purpose, the world must illustrate both good and evil. We get to know who Satan is by seeing his work through people. The intrigue in the world shows us how Satan works and criminals give us a flesh and blood representation of demons.

Generally, doing right instead of wrong brings benefits in life but that is not always the case or we would be good just for our own benefit instead of loyalty to God. Sometimes we do not see what is wrong. Sometimes we know but do it anyway. Sometimes we do right but evil still happens. That is how God wanted it to be for the test.

Horrors on earth show us the results of evils. Disasters show us that we can sometimes rescue ourselves but usually must request aid. So it is with salvation. The destruction of even the greatest of kingdoms lessens our confidence in the powers of man and lead us to put our faith in God.

IDEA # 300; THE ESTABLISHED TRUTH OF THE BIBLE: It is time that the entire world recognized that the absolute truth is the Bible. The Bible is the best-selling book in the history of the world. The teachings of the Bible are the foundation of western civilization. Which is the most prosperous civilization the world has ever known.

Jesus, the central figure of the Bible, is far and away the most influential man in history. Jesus never owned anything except his robe and incited so much hatred that he was executed in his early thirties. Yet, all the wars, conquests and, revolutions that have happened in the last two thousand years have not influenced the world as much as this one brief life even though he never held a worldly position with any more status than a carpenter.

This man from Nazareth, who shaped world events more than any other human with his short, simple life claimed to be the only begotten Son of God sent to earth to fulfill the Law and the prophets and to pay the supreme price for the world's sin. Anyone making such a claim is either telling the truth or is a lunatic. He left no room for any middle ground like claiming that he was a "prophet" or a "great teacher".

Jesus must have been telling the truth. If he wasn't, then the last two thousand years stand as an indictment of the gullibility and blind foolishness of a large segment of the human race. How could mighty western civilization achieve world predominance if it was founded on a false creed based on a lunatic claiming to be sent from God?

The teachings of Jesus and the prophets before him were often harshly condemning of the idolatry and sinfulness of the Israelites. Yet, the scriptures were treasured and preserved above all other writings. Why would they do this unless it was in fact the Word of God? The Dead Sea Scrolls leave no doubt that the Bible of today is true to the original meaning.

Jesus is exactly who he claimed to be. As hated as he was when he walked on the earth, his detractors came up with nothing to expose him as a fraud. The world's most prosperous branch of civilization got that way be having the only begotten Son of God as it's cornerstone. It is absurd to think of a book of lies being the world's bestseller for two thousand years. A book of lies is just what the Bible is if there is not a real almighty God who sent his only begotten Son to earth so that each of us can enter into an eternal covenant with him.

The simple fact that the Bible is still around today is a definite indication of it's truth. There have been numerous attempts to destroy all Bibles. The Bible has always been a problem to all kinds of evil men. Many kings and emperors who demanded absolute loyalty hated the Bible for causing the people to worship God more than them.

The Bible itself tells us that it will last forever. The Bible's promises that it will be everlasting have remained true. Despite many attempts over the centuries to destroy or discredit the Bible, it has always endured. The greatest book ever has been translated into well over a thousand languages. Many books lose their meaning when translated from one language into another but the Bible keeps it's life-changing power in any language.

Unfortunately, well-meaning Christians have sometimes given ammunition to the opponents of the Bible. It was once assumed that the sun revolved around the earth. Indeed, when one looks up at the sky it seems as if the earth is in the center and everything is moving around it. The church, which was the center of learning in the Middle Ages, taught that the earth was at the center of the universe.

The astronomical discoveries of Galileo proved the earlier theory of Copernicus that the sun, rather than the earth, was at the center. This contradicted what the church had long taught. The Catholic inquisition forced Galileo to keep silent under threat of torture but the damage had been done.

What Galileo proved with his telescope seemed to be a severe blow to Christianity. Scientific study gained new respect while veneration of the church slipped. Afterwards, more science took place independently of the church. What if the Bible was not true after all?

However, the truth is that nowhere in the Bible does it claim that the earth is the center of the universe. When it was proven wrong, it did nothing to diminish the Bible. Unfortunately, many people do not separate the Christians from the Bible itself in such matters. It is important to remember that while the Bible is perfect, the Christians are not.

The Bible, especially the Old Testament, contains much genealogy. Some Christians have attempted to calculate the approximate age of the human race by counting the number of generations back to Adam. This is another point of controversy that has unfairly discredited the Bible.

In the Seventeenth Century James Ussher, an Irish Bible scholar, made a calculation of the age of mankind based on biblical genealogy. Ussher calculated that God created Adam in 4004 B.C. It was believed that the earth and the universe were created shortly prior to this. This figure was widely accepted as the time of creation.

Of course, modern science caused headaches for Christians again. Obviously, humanity was around long before 4004 B.C. Artifacts from agricultural and nomadic settlements have been found that far pre-date this supposed time of creation. Radiocarbon dating has left no doubt.

The truth of the matter is that the genealogy in the Bible is not intended as a scientific text. It was intended for such things as confirming God's promise to Abraham to make his descendants a great nation and for prophesying the lineage of Jesus.

It may very well be that many generations are omitted in the genealogy. It was intended to show the line of descent for prophetic and historical purposes. It would be unnecessary to name every generation. King David was a descendant of the human side of Jesus and he lived about a thousand years before Jesus. Abraham lived long before David. Yet, the first verse of the New Testament calls Jesus a son of David and a son of Abraham.

Now obviously, Jesus could not be literally a son of David or Abraham. They were both long dead when Jesus was born and anyway, the Bible is clear that Joseph was Jesus' human father. But this verse serves to illustrate the real purpose of genealogy, prophetic and historical purposes. The word "son" in biblical genealogy may mean blood descendant rather than literal son. If this is the case, then it would be permissible to omit some generations from the biblical genealogy.

Another controversy revolves around the creation story in the Bible. God is said to have created the world, it's life and man in a matter of six days. Science insists that this is impossible. The world as we know it was definitely not made in six days. But before we dismiss the biblical creation story, let's consider a few relevant facts.

Scientists consider the age of the earth to be somewhere around four billion years. Plants, animals and, man are known to have come into existence gradually during the earth's history, definitely not within a mere six-day period. Is the Bible story wrong?

Remember that Moses wrote the Book of Genesis sometime around 1500 B.C. Remember also that parables are often used in the Bible when it is the most effective way of explaining something. Jesus later made wide use of parables. The creation story is not a myth but neither is it a modern scientific textbook. Just because the creation story is not what a westerner would consider a science text does not make it untrue.

I believe that the Genesis creation story is a parable. If the people of 1500 B.C., who were mostly illiterate and knew nothing of modern science were told the details about the creation of the universe and earth, it would be so far above their comprehension as to be meaningless. Such numbers as a four billion year old earth and a fifteen billion year old universe would mean nothing.

The parable of creation gave the ancients a story that they could relate to. While reading the creation story or listening to it, they could easily picture the almighty God creating the universe. If they had been given the scientific details, it would simply have bewildered Moses' contemporaries and virtually everyone else down through history.

Even our modern scientific knowledge of the creation must be far from complete. If science students of today must study hard to understand our limited knowledge of the origins of the universe and the earth, how could an ancient shepherd be expected to understand it? Unless of course, it was given to them in a parable that they were able to understand, like the ones that Jesus was later to use.

The Genesis creation story may be a parable, written for the understanding of the ancients. But in no way is it a myth. When put alongside modern science, the creation story is shown to be extremely accurate in describing the creation of the universe, the earth and, life.

Heavens and earth were created first. This would take us from the cataclysmic explosion that began the universe to the formation of a huge swirling cloud of dust and gases that was later to become our solar system. Astronomers estimate that the great explosion of energy took place some fifteen billion years ago.

The science of astronomy, which in Galileo's time seemed to point away from the Bible, has shed new light on the Bible in the Twentieth Century. The Bible has always claimed that the universe had a definite beginning, created by God. Astronomy for many years disagreed. Astronomers as a whole accepted the Steady State Theory. This theory supposes that the universe with it's multitude of vast galaxies had no beginning. It was simply always there.

But astronomers in recent years using spectroscopes attached to telescopes proved the Steady State Theory to be wrong. By observing and measuring the red shift in the spectrum also known as the Doppler effect, it has been proven that all other galaxies are moving away from our galaxy at incredible speeds. The further a galaxy is away from us, the faster it is moving away from us. Our own galaxy is hurtling through space taking us with it.

This can only mean one thing. All of the matter in the universe was together at one point when the great explosion of energy set the universe in motion. Astronomers almost all now accept the theory, known as the "Big Bang Theory", and that the universe began with this explosion. Faint radiation from the energy released by the original explosion can still be detected by modern radio telescopes.

The confirmation of the Big Bang Theory proves that the Bible was right all along. Just as Moses, inspired by God, wrote in the Book of Genesis, the universe had a definite beginning.

In order for the creation story to be written down, God must have taken Moses in a vision to show him a panorama of the creation story from the Big Bang to the creation of human beings.

The next thing to appear in the creation story after the heavens and earth is light for the earth. Our solar system got it's start as a vast cloud of gas and dust swirling in space. Later, the force of gravity condensed the cloud into the central sun and the planets. While the earth was forming into a solid whole, the sun was undergoing the process that occurs throughout the universe that forms stars out of interstellar dust and gases.

To form a star in space such as the sun, the main requirement is enough matter to get started. As the mass condenses by gravity, it creates a greater gravitational attraction leading to still more compacting, condensing force. The temperature in the center of the mass climbs higher and higher until ultimately, a thermonuclear reaction is started. A star is born and in time, light and heat radiate from it's surface when the star interior becomes hot enough to fuse small atoms into larger atoms and the unused binding energy is released in the form of heat and light.

Certain factors were necessary for the earth to form and to receive light from the sun. For all practical purposes, conditions were right for the formation of the earth only in the spiral arms of our galaxy and not in the central hub of the galaxy or the so-called globular clusters of stars surrounding the galaxy. The spiral arms contain heavier elements in the form of gas and dust.

The vast majority of the matter in the universe consists of hydrogen and helium, the two lightest elements. For the earth to form and our lives to be possible, carbon, oxygen, metals and so on are required. These heavier elements are only produced inside stars during the fusion process that produces the heat and light from these stars. The stars live out their lives and ultimately explode in a nova or supernova. The heavier matter is thrown out into space in the form of dust and some of it ultimately forms into other stars and planets.

What this means is that the sun must be at least a second-generation star and every atom in the earth as well as in your body was once a part of a star that exploded. There is abundant heavier element dust in the spiral arms of our galaxy but not much in the central hub of the galaxy or the globular clusters of stars around the outside of the galaxy. There is some controversy as to whether the planets of our solar system formed from matter directly from the sun. But either way, the earth would be likely to form only in limited areas such as the galaxy's spiral arms.

It happened just as told in the creation story. As the earth was condensing into a planet, the sun was condensing into a star. Not long after the earth took shape, it received the heat and light of the sun.

According to the biblical story, the next thing to be created was water on the earth. In it's early days, the earth was molten and hot, with numerous active volcanoes. Heat tends to accelerate chemical reactions and oxygen and hydrogen in the atmosphere of the young earth would have united to form water. At first the water existed only as water vapor in the atmosphere but as the earth cooled, the water would condense to form oceans. The earth rotated on it's axis by it's own gravity to create day and night as the story claims.

The next step is the separation of the dry land from the waters. The higher areas became land while the lower areas became bodies of water. But here we appear at first glance to run into difficulty. The Bible claims that God brought the waters together in one place. How could this possibly be when the waters of the world clearly are not together in one place? On a globe, we can see the Atlantic, Pacific, Arctic and, Indian Oceans are clearly not in one place but are widely separated. Have we discovered a flaw in the Bible?

No, it is not a flaw. As usual, the Bible turns out to be right after all. In modern times, geologists have discovered a system called plate tectonics. At one time far in the past, the dry land of the earth was together in one place. This would also mean that there were no oceans in different places on the globe, only one large mass of water just as the Book of Genesis claims.

If you look at a map of the world, you can easily see how the continents have separated and drifted apart. The ideas of plate tectonics are that the continents are on large plates that drift on the earth's mantle. The east coast of Brazil fits snugly into the west coast of Africa. The southern shore of Norway and Sweden fits into Hudson and James Bays in Canada. If we bent Central America to the west, the entire coast of the Gulf of Mexico and Caribbean Sea fits around northwest Africa from Florida to Brazil. Alaska would squeeze into the Sea of Okhotsk and Australia appears to line up with the east coast of India. Once again, modern science has verified what was in the Bible all along. The waters of the earth were indeed together in one place in the early days of the earth.

The Genesis creation story points out that the first life to be created on earth was plant life. There has been photosynthesis and the resulting oxygen in the atmosphere for about half of earth's existence. Once again, modern science has verified the biblical creation story. Not only that but the resulting oxygen produced by photosynthesis in the plants was necessary before animal life could exist on earth. How can the Bible possibly make claims which the ancients had no way of knowing but which modern science confirms unless it is really the Word of God?

After the earth began to cool and the waters condensed, the next step in the story is the creation of the lights in the sky; the moon, planets and, stars. We know that the moon and planets are probably the same age as the earth and that the stars visible at night may be older. But consider how the creation story must have been written, God giving Moses a vision of the different stages of creation. If this is the case then the early days of our world before any life on earth must have witnessed an extremely dusty and smoky atmosphere. The effects of a volcano on the atmosphere can often be seen for thousands of miles around. I remember in 1980 seeing the blue tinge to the moon from Niagara Falls caused by the eruption of Mount St. Helens in Washington State. Now, try to imagine earth covered by active volcanoes for thousands or millions of years, as it is believed the earth was in it's early days.

The stars and planets and probably the moon simply would not have been visible until the earth had aged and cooled and the dust settled down. From a point of view on earth, the celestial objects may as well not have existed. This is why the moon and stars are not mentioned at the very beginning of the creation story. They fit in a little later when they become visible from earth.

The Bible correctly points out that the first living creatures on earth lived in the sea and not on the land. The green scum in stagnant ponds is green algae and is the ancestor of chlorophyll bearing land plants. There has been a mad proliferation of life for only about the last one eighth of the earth's history. The so-called Cambrian Explosion of forms of life took place 550 million years ago. This explosion of life was based in water and not on dry land. Some of the well-known creatures of the Cambrian era are brachiopods, which were small clams, and, trilobites. In fact, much more is known about seas during the Cambrian Explosion and little about events on the land.

The Bible mentions "sea monsters" as living in the early days of life in the sea. For centuries, this must have made the creation story seem like a myth until it was discovered that in the distant past, huge dinosaurs lived and swam in the sea.

These aquatic dinosaurs along with giant pre-historic sharks could no doubt be accurately described as "sea monsters".

Birds appear early in the development of life in the biblical creation story. There were flying dinosaurs and many paleontologists believe that birds actually evolved from dinosaurs. Whether or not this is true, the fact remains that flying creatures existed on earth early on just as the Bible creation story claims.

Animal life comes next in the story. The mammals and reptiles as we know them. The creation story in the Bible is in no way a myth. With profound accuracy, it names the correct sequence of the beginning of the universe, the formation of the earth and the development of life. The statistical odds of the Bible accomplishing this just by chance is prohibitive. The people of the time the Book of Genesis was written knew very little of natural history and nothing of cosmology. This incredible creation story could not possibly have originated with human knowledge. This creation story shows the Bible to be the Word of the almighty God.

Finally in the creation story we come to man. This also stands with modern scientific discoveries. We know for sure that man, compared to other creatures is very much a latecomer on the earth.

The Genesis creation story points out a special distinction about man that is not true of any other earthly creature. The Bible states that mankind is created in the image of God, our creator. This can be readily observed. Humans are the only beings on earth with a spiritual sense. No animal or fish has shown any sign of caring about a life after the present one or paying homage to an unseen creator god.

Another way to see how man is created in God's image is to remember that God is a trinity. God consists of the Father, the Son (Jesus) and, the Holy Spirit. Mankind, unlike any other creature, is also designed as a trinity. Each individual contains a mind, a body and, a spirit.

The list in Genesis of what God created on each day until the seventh day seems to be an obvious metaphor. Unbelieving skeptics have insisted on the impossibility of the universe and the world being created in six literal days. But what if we ponder this more closely? Just what is a day anyway?

We define a day as twenty-four hours. A day for us is the time required for the earth to rotate once on it's axis, bringing a full day and night. But how could God create the earth in a day when there was no earth to measure a day by? There can be other definitions of the length of a day. It must be possible that the term "day" in Genesis 1 refers to some other, much longer definition of a day.

The planet Mars rotates in a slightly longer time than earth does. This means that a Mars day is slightly longer than an earth day. Jupiter spins very fast and has a day about half as long as an earth day. Some planets, such as Venus, may have a day longer than it's year.

The moon spins once in the time it takes to orbit the earth and has no such thing as a day relative to earth. The same side of the moon always faces earth. Relative to the sun, the moon's day would be the same as it's revolution time around earth, twenty-nine days. This can be seen in the phases in the moon as seen from earth.

If a day is defined as the period of rotation of a celestial body then a day for our galaxy is about 200 million years. Or maybe that would better be considered as a year rather than a day for a star on the fringes of the galaxy rotating around the center. Since the galaxy is not a solid whole but a cluster of celestial bodies.

What I am trying to demonstrate is that a day is a variable definition of time. A day means twenty-four hours only on earth. The days in the Genesis creation story obviously are not twenty-four hour days. It may be a metaphor or parable like those later used by Jesus. Written so those without modern scientific knowledge could understand. It could even be the definition of a day somewhere else.

It is very interesting that the Bible has split the creation story into different periods (days) because modern science has done exactly the same thing. Paleontologists tell us that life on earth as we know it came into existence in stages. These stages are divided into eras and subdivided into periods ranging from Cambrian to Quaternary. Scientists have used more time periods than the Bible to classify the different time periods in the formation of life on earth but the fact is that we can see yet another parallel between the creation story in the Bible and the knowledge of modern science.

How can the creation story in the Bible be so incredibly accurate unless the Bible is truly the Word of God? The story of earthly life in Genesis 1 may be told as a parable so that ancient man could grasp it. But look at how it compares with modern science: The Bible correctly states that the universe had a definite beginning. It was not simply always there. Long before modern geology, the Bible tells that the world's seas were once grouped together and the dry land was together. The Bible lists the development of life from plants to man in the correct order. Like modern science, the Bible divides the formation of life on earth into different eras (days). Moses, or anyone else in 1500 B.C., could not have possibly known the facts necessary to tell this story correctly. The mathematical odds of telling the story in the right order just by chance are prohibitive. How can it be doubted that the Bible is the Word of the one true God.

One of the best-known points of contention between Christians and unbelieving skeptics is the story in the Book of Genesis of Noah's Ark. It is widely doubted that there was ever a global flood and an ark that saved Noah, his family and two of each species of animal, insect and, bird.

As of yet, the flood story has not been widely accepted in the scientific community. But neither has it been disproved. Fossils of marine creatures have been found virtually everywhere on earth, including land masses. This could be the result of the flood or, it could be due to the land having once been part of the ocean floor. The flood may or may not have been related to a warming in the ice ages. We cannot dismiss the thought that the flood may have been simply a miracle from God. If God can create the universe then surely he is capable of producing a flood on earth.

Since the Bible is the Word of God and since it makes it clear that there was once a great flood, we should expect to see a few signs of this flood and sure enough we do. In Mesopotamia, which was the center of civilization both before the flood and afterward, there is a startling display of evidence for the flood. In pits dug for archeological expeditions, traces of ancient civilizations can be seen; tools, pottery shards, garbage, etc. The relics gets older as the pit gets deeper until all signs of civilization stop abruptly at a layer of clay which continues down for ten feet. The incredible thing is that the artifacts of civilization resume on the other side of the layer of clay. Other archeological excavations in the area tell the same story.

The only possible way for that layer of clay to be deposited is water and lots of it. This layer of clay may very well be our proof of the great flood. The civilization before the flood appeared as reasonably advanced. One of the major differences in artifacts from before the flood and after the flood is that pottery from before appears to be hand-shaped while that after the flood was shaped on a wheel. Apparently, the antediluvians had not invented the potter's wheel.

There is non-biblical confirmation of the great flood. The Babylonian Epic of Gilgamesh is a story of a destroying flood that killed all except the people in an ark that floated on the waters. Babylon was near the location of the archeological excavations that discovered the layer of clay. (I would like the credit the excellent book "The Bible as History" by Werner Keller, published in 1980 by Hodder and Stoughton and reprinted by Bantam Books)

The Bible claims that the flood covered the whole world. Yet, the layer of clay only shows up around Mesopotamia. It is easier to build a case for a localized Mesopotamian flood than it is for a global flood.

Remember that in the days before the flood, Mesopotamia may well have been the whole world from a point of view of human civilization. It would be totally unnecessary for God to flood the uninhabited areas of the earth or those areas not included in the development of the Bible story. If civilization existed only in Mesopotamia and Mesopotamia was flooded, from a practical point of view the flood could be said to have covered the whole world. Unfortunately, we do not at this time have all of the necessary data from archeology and natural history to determine this with certainty. You can be sure, however, that the Bible will always ring true.

Many of the misunderstandings concerning the Bible may very well originate with the errors of sincere but mistaken Christians. Creationists have supposed that fossils are the petrified bodies of creatures killed in the flood and that the fossils were deposited by the waters. Nowhere is this in the Bible, this assumption is clearly wrong and will do nothing except unjustly discredit the Bible.

While the Bible contains over-abundant evidence of it's truth, in some cases God just sends us his message and expects us to believe it. After all, he is God. Subsequent scientific discoveries often show that what was assumed to be merely a myth may be true after all. The Book of Exodus mentions the Nile River being turned to blood. Skeptics doubted for centuries. But modern environmental science now tells us that there is a red species of algae that gives water a blood red color when polluting the water.

A favorite question of unbelieving skeptics of the Bible is how Noah's Ark could possibly hold two of every creature on earth along with enough food and fresh drinking water for the journey. My guess is that it would require an ark at least the size of Rhode Island. This would still leave the problem of capturing the animals in the first place and ensuring that they get along with each other on the ark. It is surely a job for a very skillful zookeeper. God commanded Noah that for certain creatures, he was to take not one pair but seven pairs.

However, since we can make a case for a localized flood in Mesopotamia, there is a strong possibility that it would only be necessary to accommodate those creatures native to Mesopotamia. There were species found in the region in the past that are not present now. The Book of Daniel mentions that much later, Darius kept lions for hunting. Most of the creatures in Noah's Ark may have been farm and domesticated animals and local species that would have gone extinct in the flood.

It must be recognized that anything said about the creation of earth and it's life, the flood story or the early days of the planet cannot at this point be conclusively proven. We have plenty of reason to believe that the Bible is the Word of God and we should therefore believe what it says.

The differences between the Christian community and the secular world usually begin when well meaning but mistaken Christians try to add their own ideas onto biblical revelation. I have presented what I believe to be a very probable scenario for the creation story and the flood. However, my personal ideas are not equal to the revelations in the Bible. If my ideas as presented here prove to be inaccurate then it is me, not God that was in error.

The Theory of Evolution is another point of controversy between Christians and non-Christians. Some claim that life could have "evolved" instead of having been created by God.

The principle of natural selection, upon which the Theory of Evolution is based, can be observed in nature. One of the best examples is seen in crop dusting, this is the spraying crops with pesticides to kill destructive insects. If it turns out that a small minority of the targeted pests are by chance immune to the pesticide being used or at least better able to resist it's effects, then eventually all of the targeted pests in the area will be born with the ability to resist the pesticide.

This is simply because those pests with the innate ability to resist the pesticide are far more likely to survive and reproduce and thus pass on this ability to it's offspring through genetics. Those without this natural resistance will probably be killed, thereby removing their genes from the genetic pool.

Living organisms have the built-in ability to adapt to their environment, both in the short-term and over successive generations. If we do heavy physical labor on a regular basis, our bodies become stronger to adapt to the labor. Adaptation means the development of characteristics helpful in a particular environment. Over many generations, Negroes have developed a large quantity of pigment in their skin as protection against the hot African sun. Many Eskimos are short and stocky to conserve heat.

An excellent illustration of adaptation is breeding of domestic animals, whether dogs or race horses, desirable offspring are bred by matching the required genes in the parents. The only difference between breeding and natural selection is that natural selection favors those parents having the necessary traits to survive and reproduce just by chance. While breeding, in contrast, is arranged by the animals' owners.

Evolutionists take the principles of adaptation and natural selection to be the foundation of the Theory of Evolution. Obviously, these principles are very important in the natural world.

The great fault in the Theory of Evolution is that it cannot adequately explain how life on earth got started in the first place. The concept of natural selection as seen in crop dusting and adaptation have obvious truth. But the evolutionary idea of life arising from random collisions of atoms simply does not fit the facts. The Theory of Evolution can explain the survival of the fittest but cannot explain the arrival of the fittest. It is understandable how life can produce other life (biogenesis), but the natural physical forces and laws would not make it possible for life to come about from non-living matter (abiogenesis) by the processes of nature alone.

The old arguments against the creation of life by random natural processes are as true as ever. Even the simplest living organism is extremely complex. Atoms react to form molecules but the evolutionary idea of molecules combining with other molecules to form complex structures runs into difficulty. Molecules form naturally from random collisions of atoms but once formed, molecules show little tendency to form sophisticated structures with other molecules that are necessary for life. Most molecules, water molecules for example, end up grouped together with other identical molecules.

The Second Law of Thermodynamics, which is far more scientifically established than the Theory of Evolution, states that in the normal course of nature the tendency is toward chaos and disorder rather than structure and order. Natural forces do not build complex structures out of simple structures. Instead the complex tends to be broken down into the simple. Nature seems to prefer disorder to order in situations such as those in which evolutionists claim life arose from inanimate matter. This natural tendency toward disorder is called entropy. The universe is an extremely orderly place, but the basic forces of nature, which maintain this order, would not apply to random collisions of molecules and atoms building a structure leading to a living cell.

The difficulties for the Theory of Evolution begin when the theory is taken too far. Adaptation and natural selection clearly occur in nature. But when the theory is taken much further than this, it ceases to fit the facts. The Theory of Evolution is wrong, not because it has no truth in it, but because the evolutionists have tried to take it too far. The fact that species have the ability to adapt to their environment has been blown up to try to explain how life originated in the first place.

Evolutionists point to similarities in the structures of different creatures, apes and man for example, and take it as proof that the creatures have common ancestors. But can this really be considered as proof? Man and other creatures are designed as they are for convenience and practicality. Apes and man have similar designs because it is a practical and useful form for both.

If the Theory of Evolution is really as true as claimed then every creature should be continuously improving it's fitness for it's environment and thus changing. If changes occur over the generations then it naturally follows that those creatures with the shortest life spans and therefore the most generations should evolve the fastest. Yet, this is not the case. On many occasions in pre-historic times tens of millions of years ago, insects such as ants were trapped and covered in amber from pine trees. The insects were preserved as the amber hardened and were found in modern times. According to evolutionary theory, insects with a short life span and thus, more generations should be very drastically changed over millions of years.

But something must have gone very wrong with evolutionary plans because the ants were the same as modern ants. There were no doubts that the specimens were tens of millions of years old.

Ever since the days of Louis Pasteur, scientists have been trying to create life in the laboratory. Not one of the many experiments have succeeded in creating any form of life, not even on a microscopic level. Scientists know the conditions of the time periods in which evolutionists believe life on earth originated and can recreate those conditions in the laboratory. A wide variety of conditions with various elements and compounds present in the earth, air and, water. All ingredients necessary for life have been included in such experiments.

Evolutionists believe that the beginnings of life on earth occurred naturally from matter in the environment. If life, even in it's simplest forms, can occur so readily in nature then why could brilliant scientists experimenting with a wide variety of conditions not succeed in creating life in the laboratory? The amino acids and other materials for life can be found but no actual life ever appears.

The only explanation that I can think of is that life did not originate as a result of natural processes. It originated from God.

The development of life on earth shows signs not of following evolutionary principles but of being guided by God. Modern science now tells us what the Bible has claimed all along, that animal life began in the oceans. One of the great mysteries of evolution is why creatures would leave the water for dry land. An aquatic creature would develop no lungs or legs for life on land. Even if, by some wild stretch of the imagination, an aquatic creature managed to appear that could breathe air, it's chances of survival would be far less, not greater on land. When a whale gets beached it does not increase it's chances of survival, it faces almost certain death. Life could only have moved from sea to land if God had wanted it so.

The greatest challenge to creatures in the wilderness is finding enough food. Keeping this in mind, we notice another mystery that does not fit the facts in evolutionary theory. It does not make evolutionary sense for cold-blooded animals to evolve into warm-blooded animals. A cold-blooded animal requires less food, less variety of food and fewer vitamins than warm-blooded mammals. If some cold-blooded creature many millions of years ago had been somehow born with warmer than usual blood, it would require more food and a wider variety of food. This would increase it's difficulty in finding sufficient food and therefore it's chances of dying of starvation.

The emergence of warm-blooded creatures from cold-blooded definitely happened but it is also most certainly opposed to the principles of survival of the fittest. It is difficult to explain occurrences such as this and how vertebrates came about from invertebrates, unless, of course, God wanted it to be this way.

Man is one creature that refuses to exist as evolution supposes he should. A car is an excellent example of mechanical principles. An airplane is a fine example of aerodynamic principles. But a human being is a very poor example of evolutionary principles.

In the evolutionary game, the goal for a creature is to survive and pass it's genes on through reproduction. Natural selection will have provided the creature with adequate strength, intelligence and, senses if it is fortunate. It would make no evolutionary sense for a creature to have excessive abilities that would not contribute to survival and the passing on of genes.

It is fine for human beings to possess enough intelligence to hunt, fish and, farm. But the ability of the human race to create masterpieces of literature, art, mathematics, philosophy and, religion are totally unnecessary for evolutionary survival. Too much intelligence can be not a bonus but a positive hazard. When man developed technical abilities, he used much of it to develop weapons to kill other people.

If man is the most highly evolved creature on earth as evolutionists claim, then humans should be the finest example of evolutionary principles of any creature.

But once again, we have a puzzle that defies evolutionary theory. If we consider man in light of evolution, the first thing that does not fit the facts is the human brain, which has a capacity for intelligence far beyond necessity for survival and reproduction that is the objective in the Theory of Evolution. This is in mysterious contrast to other creatures.

Religion has invariably been a vital part of human existence since the earliest pre-historic times. The forms that religion has taken have varied widely but a powerful attraction to and curiosity about the supernatural is a well-established aspect of humanity. The most influential figures in human history have been religious figures just as the best-selling books have been religious books.

If the Theory of Evolution is true then there is no such thing as the supernatural, nothing beyond what we can detect with our senses. Man should be the best example of evolutionary principles since we are supposedly the highest level of evolving life.

Why then, must we ask, is humanity so obsessed with the supernatural? No evidence has been shown that any creature besides man has any interest whatsoever in anything supernatural. It seems to interest only humans and has done so throughout the existence of humanity on earth. If evolution is true, what possible reason could there be for a consuming obsession for something that does not exist to evolve in man, which is supposedly the highest example of evolution? If the great goal in evolution is simply to survive, reproduce and, pass on one's genes, then religion and anything else supernatural serves no useful purpose whatsoever.

Unlike evolutionary theory, the Bible has a fitting explanation for man. The Bible claims that God created man to have dominion over the other creatures and gave man a spiritual sense unlike the other creatures. If humans are merely collections of molecules operating by the logical forces of nature, then why have we evolved emotions? If we are logical creatures resulting from evolutionary principles then our capacity to feel emotions is as nonsensical as our attraction to the supernatural.

Our emotions, particularly the negative side of our emotional range, must be considered as more detrimental than beneficial to survival. If we are evolved collections of molecules than why should we feel happiness or sadness, love or hate? Different situations are merely different collections of atoms, is it not a breach of evolutionary logic for a collection of atoms and molecules to feel joy, anger or, disgust over the arrangement of the atoms around it?

No theory can explain and account for human nature like the Bible can. The Bible tells us that God created man in his own image with a spiritual sense but also with the potential for evil. This logically fits with what a human being is, not the Theory of Evolution. Mankind, along with other creatures, has the ability to adapt to the environment as the evolutionists claim. But the Theory of Evolution loses it's legitimacy when attempts are made to use the theory to explain the origins of life. The Bible, however, gives a perfectly accurate description of what a human being is and what his nature is.

The Theory of Evolution and it's idea that life on earth originated from natural processes cannot be true. One of the reasons that the theory is so popular is that it makes no moral demands on anyone. If we are just collections of atoms, products of evolution, then there is absolutely no reason for any kind of moral restraint.

The goal in evolution is to survive and reproduce. If we can increase our chances of survival by taking something belonging to someone else, then why not do so? If we can pass along more of our genes by killing someone and taking their mate, then what should stop us? If we are merely products of evolution as the animals are, then what possible reason is there for us to live differently than they? If a man is only a collection of molecules then what is wrong with murder? If this life is all there is and we have no God to answer to, then why should we not grasp all we can for ourselves now? This is the essence of evolution and maybe it can explain why the past one hundred and forty years or so have been such a chaotic and deadly time in the world.

If the Theory of Evolution were true it would not be a noble scientific theory, it would be the brutal law of the jungle. What even non-Christians call evil behavior would be perfectly acceptable if evolution was the truth.

A married man who got six other women pregnant may be frowned upon by society, at least some of the time. But he would be doing just great according to the Theory of Evolution by passing along more of his genes. A person with absolutely no compassion for anyone aside from himself would also be frowned on occasionally. Yet this attitude would be perfectly in harmony with evolutionary principles.

If the Theory of Evolution was really true, then the best thing we could do to ensure a better future for humanity is to breed human beings as we do dogs and horses. Why not sterilize the least intelligent and capable people to prevent them from passing along their inferior genes and pay the most intelligent people to produce more offspring?

Evolution is a brutal idea, favoring the fittest and devouring the rest. If this theory were true then the ultimate reality would be simple randomness and survival. The theory had been popularized and had gained wide acceptance by the early Twentieth Century. The Theory of Evolution made killing much easier on the conscience, modern technology made possible fearsome weapons and, the result has been a very deadly era. It is possible that more people have been killed in the last hundred years by other human beings than in all previous centuries combined. I cannot help but wonder to what extent this vast destruction can be pinned on a wide acceptance of the Theory of Evolution.

Still, no doubt can seriously be held that a sizable portion of the reason for the popularity of the Theory of Evolution is that many people do not want to believe in God. Though they may not fully understand the theory and it's shortcomings, most people are aware that the Theory of Evolution makes no moral demands on anyone. In this theory, we are just a random cosmic accident with no god to answer to. So, the reasoning goes, there is nothing much wrong with doing whatever we want to do. Any conscience or innate morality we may have must be just the residue of now outdated and discredited religious beliefs. Or at least so the theory goes.

The controversy in the U.S. over whether the education system should emphasize the creation story or evolution in schools has been raging for years going back to the Scopes Trial in 1925 in which John Scopes, a biology teacher, was convicted of teaching evolution in the classroom. The conviction came after the Tennessee legislature had passed a penal statute earlier in the year making it illegal to teach in schools anything that denies the divine creation of mankind.

However, such laws favoring biblical creation were later repealed and in 1963, Madalyn Murray O'Hair won a suit against the city of Baltimore resulting in the decision of the U.S. Supreme Court to outlaw Bible reading and prayer in public schools. In the education controversy, the evolutionists have gotten their way.

It is difficult to see the justice of the Bible being banned from public schools while the Theory of Evolution is permissible. Not only does the theory not fit the facts when it is used to explain how life originated on earth, it is just as much an article of faith as the Bible is. Evolution is a concept that is unproven and would dominate an individual's view of the meaning of life if it were accepted. In the basic sense, evolution can be described as a religion. It requires at least as much faith to believe that life arose on earth by random collisions of atoms as it does to believe that God created life on earth.

If there is no God then the entire universe as well as the earth and it's life must have come about through the random coincidence of natural forces. Everything that the universe is, everything it contains, every event that has ever happened or ever will happen must be the result of pure chance.

On the other hand, if there was a God that created the universe and planned living things to be a part of his creation, we could expect to see an extremely complex universe with many factors coming together to make life possible. And everything about the earth and the universe indicates that this is just the way it is. The conditions making life possible, even aside from the miracle of life itself, are not the random result of natural forces but fortunate blessings from God.

In the course of our day-to-day existence, most of us give little consideration to the myriad of factors that have made it possible for us to have life at all. There is a vast number of conditions on earth and in the properties of matter and energy in the universe which if altered would have made our existence impossible.

It is my belief that the conditions that made the human race possible were so numerous and complex that it could not have been this way simply by chance. The odds of everything being right for life and humanity to exist occurring by random chance are so minute as to be virtually incalculable. The only fitting explanation is that the vital conditions came about because the God that created the universe wanted it so.

We do not know everything about the Big Bang, the cataclysmic explosion some fifteen billion years ago that began the universe. We do know that it is extremely difficult to explain. Just what could cause such a vast explosion of energy?

The Big Bang must have been an explosion of energy and not of matter. Consider the objects known as black holes. Every celestial body has what is called an "escape velocity", which is the speed at which an object can escape the body's gravity. The earth's escape velocity is seven miles per second. Meaning that if we could shoot an arrow into the sky at this speed, it would escape earth's gravity and continue on into space. But at anything less than this speed, it would reach a certain height and then fall back down due to gravity.

The more massive a celestial body is, the greater is it's escape velocity. Density is also a factor in escape velocity. If two bodies are of equal mass, the one highest in density will have the highest escape velocity as long as the starting point on both bodies is the surface. The giant planet Jupiter has an escape velocity much greater than the earth's while the moon's escape velocity is much less than the earth's.

No object in the universe can travel faster than the speed of light, 186,282 miles per second. This is the speed at which light and other electromagnetic radiation travels. According to the Theory of Relativity, extremely high velocities play bizarre tricks on matter. One of which is that an object's mass greatly increases as it approaches the speed of light. At the speed of light, the object would have infinite mass. To accelerate it to a higher speed would require an infinite force, which is of course impossible. This leaves the speed of light as the universal speed limit for everything, matter and energy alike.

Now suppose that a body is so dense and so massive that it's escape velocity exceeds the speed of light. Nothing, not even light, would be able to escape. Such a body is known as a black hole. Any matter approaching close to the black hole would be pulled in and would disappear forever. Of course, we cannot see black holes because no light can be reflected from their surfaces although radiation given off by matter captured by the black hole's gravity can be detected.

Now let's go back to the Big Bang. If we keep in mind how a black hole operates, it becomes obvious that the Big Bang was an explosion of pure energy and not of matter. Matter, having gravity, could not have been present at the Big Bang. If even a tiny fraction of the matter in the universe had been in close quarters, it would not have created a Big Bang. It would for an incredibly powerful black hole and the universe would consist of this black hole and nothing else.

Albert Einstein's Theory of Relativity shows that matter and energy is interchangeable. When we burn fuel, we convert some of the mass into energy. A nuclear reactor or bomb converts mass into energy even more directly. Given the right conditions, energy can also be converted into matter. Therefore, the matter in the universe was converted from the energy in the Big Bang.

It simply cannot be explained by scientific means what would cause such an explosion of pure energy of unimaginable magnitude. It would be far easier to explain an explosion of matter than one of energy.

The creation of the material universe from the cataclysmic explosion would also be complex and not easy to explain. What would convert the energy radiating from the explosion into the matter necessary to construct the universe? The energy of the new universe could not condense or be transformed into matter too soon or it's own gravity would pull it back into a massive black hole.

Matter and energy can be interchanged according to Einstein's theory but a small amount of matter is interchangeable with a vast amount of energy, as in the atomic bomb. Where would such an unimaginable quantity of pure energy that it would require to transform into the matter of the universe come from? It is mind-boggling and science has scarcely tried to explain it.

If we consider the concept widely believed by astrophysicists that the Big Bang brought into existence vast amounts of both matter and antimatter then the large amount of energy released in the Big bang becomes even more difficult to explain because there was therefore much more matter than we see in the universe today. Supposedly, there was both matter and antimatter created from the energy explosion but slightly more matter. The matter and antimatter is supposed to have annihilated each other in a great holocaust leaving the little bit of matter that had no corresponding antimatter intact. This little bit of leftover matter can be seen today, it is what the universe is made of. According to another theory about matter and antimatter, the matter that we know is more stable and so most of the original antimatter created from the energy from the Big Bang no longer exists.

What all of this means is that the Big Bang must have involved an almost infinite explosion of energy. This leaves us with a puzzling question. Where could all of this energy have come from?

One way of answering this question is to speculate that the universe is born and dies in cycles. Maybe the universe was born in a big bang and extended outwards for billions upon billions of years until it's gravity slowed the expansion and the galaxies fall back together again. The universe would therefore end with all the galaxies colliding in a cataclysmic explosion. But the explosion sets the whole process in motion again and a new universe is born. It, of course, shares the fate of the universe before it and still another universe would come into being.

(By the way, this book is being written in the U.S. and by "billion", I mean a thousand million. If you are outside North America, please substitute "milliard" for billion.)

However, this speculation still does not answer our question. We have absolutely no evidence of any universes prior to our own. Even if there were previous universes and the universe was born and dies in cycles, the process must have begun at some point. There must have been an original Big Bang. We are led right back to the original question about the Big Bang that began the universe. Where did the almost infinite amount of energy come from?

An answer to this question is that the Bible is the truth, God really exists and he provided the almost infinite amount of energy that it took to start the universe. Many times, discoveries have confirmed claims that were in the Bible all along. Someday, the very origin of the universe will no longer be doubted to be, just as the Bible claims, a creation of God. There is simply no natural explanation for the almost infinite amount of energy that brought about the Big Bang.

Just as there is no way to account for the beginning of life on earth by natural forces and processes, neither is there a plausible explanation for the origin of the universe itself. It seems to me that the only fact-filling account is the creation story in the Bible.

The atom, which is the building block of the universe, is a work of sheer genius. The sub-atomic particles that compose the atom can be assembled into over a hundred different kinds of atoms. This is essential for life. If only one kind of atom or only a few kinds were possible, life could not exist.

Fortunately, some of the sub-atomic particles are electrically charged. Protons, which are in the atomic nucleus, have a positive charge while electrons that orbit the nucleus have a negative charge. God designed the atom to be electrically neutral under normal circumstances and also for the electron shell of each atom to prevent other atoms, also with negatively charged electron shells, from passing through it because the atom is mostly empty space.

When two electrons with the same negative electrical charge are in close proximity, they repel each other. If this were not the case, then atoms would blur into each other when in contact and matter of any form, including life, would be impossible.

We could have no matter as we know it and therefore no life without the nuclear force that holds the atomic nucleus together. It is this force, also known as binding energy, which gives off the power as a nucleus is split in a reactor or a bomb during fission. When we first consider the structure of the atom, it seems bewildering how the protons in the nucleus can remain grouped together if like charges are supposed to repel. But without the protons close together in the nucleus, matter would be impossible and so would life.

Fortunately, or rather by God's design, there is the nuclear force that is much stronger than the electromagnetic force but is effective only for extremely short distances so that the force does not interfere with the atom's orbiting electrons.

Not only are atoms the building blocks of the universe, they are also it's batteries. Every star in the universe generates it's vast power by fusion. In these stellar nuclear reactors, four hydrogen atoms are fused together to form one helium atom and the leftover mass is converted to energy that reaches us as sunlight or starlight.

Maybe ninety percent of the matter in the universe is hydrogen. We could say that this is the ash left over from the Big Bang. Hydrogen, which is the simplest of all atoms, formed out of the countless particles that condensed from the energy released in the Big Bang. When we really stop to think about it, we can begin to visualize God's infinite mind planning and creating the universe.

How very fortunate we are that matter and energy are interchangeable. If they were not, the universe could neither function nor could have gotten started in the first place.

As it turns out, there is just about the right amount of matter for the universe. Had the cosmos contained too much matter, it would have collapsed back in on itself billions of years ago due to the powerful gravitational attraction that would have come with increased mass. As it stands now, there is enough matter to build the universe but not so much to introduce enough gravity into the early universe to cause it to draw all the matter back together again. Of course, this would make our lives impossible.

How can we explain the basic forces of nature by which the universe operates? Science assumes that the forces of gravitation, electromagnetism and the nuclear binding force just exist without explaining their origin.

In these half dozen or so basic forces, we have another factor without which the universe would not exist. What could cause the forces that work so wonderfully together to make our lives possible? The fact that matter exists is hard enough to explain. When we add the fact of the necessary forces working in just the right way, it becomes obvious that the universe was planned and set in motion by an all-powerful creator who wanted the universe to act as it does.

Our cosmos is an awesome mechanism of intricate design. We take the basic nature of our surroundings for granted without realizing that if it were any different, life would never have existed. There is a stable foundation of basic laws and forces in our universe that makes possible us and all that we know. It makes the idea of all this being a cosmic accident ever more ludicrous, the existence of a great creator God all the more certain.

The fact that there are three spatial dimensions and one time dimension is not an accident of nature. It is God's design. Any fewer dimensions and life would not be possible. It is difficult to imagine the universe if there were more than four dimensions. Our four dimensions are just right. Without the time dimension, matter could exist but not energy. Motion would not be possible without the time dimension and so neither we, nor our universe, would exist. Every facet of the cosmos reflects God's supreme wisdom.

One more example of this great wisdom is heat. God designed matter so that it is possible not only for an object to possess kinetic energy of motion but also for individual atoms and molecules to move with kinetic energy. This makes heat possible without which our universe would be very different.

Related to heat is state of matter. There are three states of matter: solid, liquid and, gas. The higher the temperature, the greater is the kinetic energy and movement of the molecules and so, the looser the bond between molecules. A gas forms when the bonding is loose, a solid when it is tight and, a liquid in between. One of the best examples is water in it's three forms that depend on temperature: ice, liquid water and, water vapor.

Even such a simple thing as friction would make life in our form impossible if it were absent. Solid matter, unless intensely polished, is not smooth but rather it's surface consists of microscopic mountains and valleys. Friction is the collision of these tiny mountains on the surfaces of two objects in contact. Friction is essential for walking or crawling, which is why it is easier to walk on cement than on a patch of ice. It would be necessary for human, mammal, reptile and, amphibian life to exist in very different forms were it not for friction.

God planned for electromagnetic radiation, the conveyer of energy throughout the universe, to travel through empty space at a certain speed that does not vary regardless of conditions or the wavelength of the radiation. Let's take a close look at this for a moment. Suppose that there were factors that would cause the speed of electromagnetic radiation to vary significantly. When a moving object of constant mass has more kinetic energy, it moves faster. However, an electromagnetic wave moves at the same speed regardless of it's energy level. But, what if it's speed did vary?

The natural law governing the speed of electromagnetic radiation is another of those rarely appreciated blessings that are essential for life to exist on the earth. If the steady flow of electromagnetic energy to earth from the sun and to a much lesser extent, the stars, were subject to variation in it's speed in traveling through space, sooner or later life on earth would be either broiled or frozen by either a great excess or lapse of radiation reaching our planet.

The visible light to which our eyes are sensitive is only a tiny portion of the entire electromagnetic spectrum that ranges from long radio waves to cosmic rays. But this relatively narrow wave band on the entire spectrum is ideal for gathering information for the purpose of life on earth. Had man, before the advent of modern civilization, had organs sensitive to any radiation except light, he would have gotten little information of practical value from his earthly surroundings but a barrage of radiation from space. Had our eyes been able to receive all or a large part of the electromagnetic spectrum, we would be utterly overwhelmed by sensory overload if we dared to open our eyes.

By what is known of astronomy, the one celestial body that stands out in sharp contrast to the rest of the known universe is our earth. Our planet gives every indication of being custom made for life by God. Compared to other bodies in space, earth is like a sheltered oasis in a hostile and forbidding universe.

The sun and other stars are blazing nuclear furnaces that would vaporize any life within millions of miles. The space between galaxies and even between stars within galaxies is empty and so cold that virtually all of the molecular motion stops.

Unlike earth, the other planets in the solar system are not very welcoming. Mercury is hot enough to melt lead. Venus is covered by dense clouds that cause a massive greenhouse effect and has enough atmospheric pressure to crush exploring space probes. Mars is a cold desert with an unsuitable atmosphere. Jupiter has crushing gravity, violent weather and a poisonous atmosphere of methane and ammonia. Saturn is similar to Jupiter but colder. Uranus, Neptune and, Pluto are too freezing cold to even imagine the existence of life. The asteroids between Mars and Jupiter are barren rocks with little or no atmosphere.

The earth, however, seems to be made especially for life. It has water, oxygen in the atmosphere and an ozone layer high in the atmosphere to keep out deadly ultraviolet radiation from the sun. Unlike some of the other planets, the earth rotates as it orbits the sun in a convenient interval of time. This prevents one side of the earth from getting too hot while the other side gets too cold.

Fortunately, the earth's surface is different. Some areas are water, some are ice, some desert and, some woodland. This causes the earth to heat unevenly. The uneven heating of the earth's surface produces weather. Our weather supplies water to life far away from bodies of water. The evaporation process acts as a natural distiller to remove the salt in the sea that would otherwise prevent plants from growing in the soil if it fell with the rain.

The variation of the level of the earth's surface is just enough to permit the smooth operation of the weather cycles necessary for life. If there were too little variation in the levels of the earth's surface, water would cover the entire surface. If there were too much variation, air would pile up in the lower areas leaving much highland area uninhabitable and the oceans would cover less surface area thus providing less water vapor to the atmosphere meaning that there would be less water falling as rain to sustain life on land.

Our atmosphere seems to be a miracle in itself. It shields us from deadly radiation and meteors from space. It breaks down dead organic matter so the nutrients can return to the biosphere. Fortunately, air is transparent instead of opaque so that we can see through it and oxygen and nitrogen do not react chemically with each other, making air useless for breathing.

Can it be considered just our good fortune or random luck that plants breathe carbon dioxide and give off oxygen while animals do the opposite, taking in oxygen and exhaling carbon dioxide? Without this convenient respiratory cooperation between plants and animals the atmosphere may have been unable to sustain life by now.

There are so many factors that just happen to be in favor of life without which life could not exist today. How can we imagine that this could have possibly come about without a divine creator? It is often not until man-made pollution threatens the environment that we really begin to realize how that environment, against all odds of random chance, was in all probability created to sustain life. Not only is life a miracle but life's supporting network is also a miracle.

If there is a vivid example of God's design in nature aside from life, it is water. A molecule of water is composed of two atoms of hydrogen and one of oxygen. When water formed early in earth's history, it took only half as many atoms of oxygen as hydrogen to form the earth's water. This used up virtually all of the hydrogen gas in the early atmosphere but left plenty of oxygen free for the atmosphere. We were dependent on this oxygen to sustain life.

Now suppose that water had a different chemical formula requiring more than one atom of oxygen for each molecule. This would leave little or no oxygen in the atmosphere that we need for life. There is no rational way that this and so many other vital factors of life can be credited to anything but the planning of the creator God.

Water itself is a simple compound but is indispensable to life. This vital liquid shows every indication of God's design. Most compounds of hydrogen are poisonous to life, for some reason water is not. Water has the capability to absorb large amounts of oxygen at low temperatures to sustain ocean life. In contrast to the rest of nature, water expands and becomes less dense as it freezes. If this were not the case, marine life in bodies of fresh water would be unable to exist because ice would sink causing the water to freeze solid. Water is used by life only in it's liquid form. Fortunately, water remains liquid over the entire range necessary for life.

What would the world be like if the sediment that collects on the bottom of bodies of water was lighter than water instead of heavier? It would float on the water's surface, covering the body of water, blocking sunlight from the water and preventing evaporation. We can see that even a factor as simple as this, if altered, would make earthly life impossible.

What do you think of dust? The air holds countless tiny particles that slowly settle to earth. Particles of dust find their way into the smallest cracks and is usually considered as nothing more than a nuisance that must be cleaned up periodically.

But something that we consider as merely a nuisance is vital to God's creation. Dust is necessary as condensation nuclei. That is, particles of dust serve as bases for water vapor to condense into water droplets when the vapor reaches a height cool enough to induce condensation. Thus, we have clouds forming. Without clouds, there could be no rain, no plant life, no fresh water and, no life except in the oceans. Without tiny particles of dust floating in the air, life on land could not exist.

In the opinion of this author, a multitude of factors exist that would have made life impossible if any one of them were altered. Yet, every one of these factors has turned out to be in favor of life. It appears to me to be far more logical to credit this to God than to random chance.

We could continue for a long time with a list of how fortunate we are to have so many factors vital to life in our favor when they could have easily tilted the other way. Hidden in the earth are all the resources necessary for man to build great civilizations. There is a layer of topsoil without which plants could not grow. Fortunately, the dinosaurs were removed from the earth long before the advent of man. Humans would have great difficulty getting permanent settlements started with hungry carnivorous dinosaurs roaming around.

Remember that the main reason for arguments against the truth of the Bible is the moral demands that are made by the Bible. Someone living a sinful life does not want to hear what a sinner they are. They do not want the Bible to be true. They would much prefer that life had originated by random collisions of atoms so that there would be no moral demands.

Many unchristian things have been done in the name of Christianity, the crusades and the inquisition of history and the false cults and priest scandals of today. But it is a dire mistake to judge God or the Bible by the acts of everyone claiming to be a Christian. Even sincere Christians are still imperfect human beings.

There is a basic difference between a religion and an ideology. An ideology such as communism or democracy cannot be true or false and cannot be readily judged as such. An ideology can only be evaluated by how well it works or does not work. A religion, however, can only be accurately judged as either true or false. The true religion should show itself in the conduct of it's believers. But, it must be remembered that bad conduct in so-called Christians cannot disprove the existence of God or the truth of the Bible. Poor behavior in Christians comes not from following the Bible but from their not following the Bible.

Many people have been turned off to Christianity by having it forced on them in their younger years. It can be easy for a potential Christian to be dismayed by some Christian groups adding their message onto the Bible or other groups softening and watering down the message of the Bible.

However, any mistakes, hypocrisy and, conflicts between churches even though it may make Christianity look bad at times, in no way disproves the Bible. When the Bible is really followed and applied, the evils are rooted out and the fruits of salvation become visible. But it must be remembered that while God and the Bible, are perfect, the believers are not.

0-595-31163-6